# ADSORPTIVE REMOVAL OF MANGANESE, ARSENIC AND IRON FROM GROUNDWATER

# Adsorptive Removal of Manganese, Arsenic and Iron from Groundwater

DISSERTATION
Submitted in fulfillment of the requirement of
the Academic Board of Wageningen University and
the Academic Board of the UNESCO-IHE Institute for Water Education
for the Degree of DOCTOR
to be defended in public
on Wednesday, 28 October 2009 at 15:30 hrs
in Delft, The Netherlands

*by*

## RICHARD BUAMAH
*born in Kumasi, Ghana*

*CRC Press/Balkema is an imprint of the Taylor & Francis Group, an informa business*

© 2009, Richard Buamah

Published by:
CRC Press/Balkema
PO Box 447, 2300 AK Leiden, The Netherlands
e-mail: Pub.NL@taylorandfrancis.com
www.crcpress.com - www.taylorandfrancis.co.uk - www.ba.balkema.nl

ISBN 978-0-415-57379-5 (Taylor & Francis Group)
ISBN 978-90-8585-526-2 (Wageningen Universiy)

# CONTENTS

Acknowledgements ......................................................................................... ix

Abstract ......................................................................................................... xi

**1    Introduction** ....................................................................................... **1**
    1.1    Water consumption, resources and quality ............................................ 2
            1.1.1    Groundwater use and quality in Ghana .................................. 4
            1.1.2    Groundwater use and quality in the Netherlands .................. 5
            1.1.3    Groundwater quality problems................................................ 6
    1.2    Manganese, arsenic and iron removal methods ..................................... 6
    1.3    Origin, mobilization and chemistry of manganese, arsenic and iron ........ 11
            1.3.1    Manganese in groundwater: reduction and oxidation ............. 11
            1.3.2    Arsenic in groundwater ........................................................ 18
             1.3.3    Iron in groundwater: reduction and oxidation ....................... 24
    1.4    Theoretical background of adsorptive processes ................................... 27
            1.4.1    Equilibra ............................................................................... 27
            1.4.2    Forces and energetics of adsorption ...................................... 27
            1.4.3    Adsorbate – solvent properties ............................................... 28
            1.4.4    Adsorption isotherm .............................................................. 29
    1.5    Need for research ................................................................................. 29
    1.6    Research objectives .............................................................................. 30
    1.7    Outline of the thesis ............................................................................. 31
    1.8    References ............................................................................................ 31

**2    Presence of manganese, iron and arsenic in the groundwater within
the gold-belt zone of Ghana** ............................................................... **39**
    2.1    Introduction ......................................................................................... 40
            2.1.1    Sources of manganese, iron and arsenic in groundwater ........... 41
            2.1.2    Simplified geology of Ghana ................................................. 41
            2.1.3    The gold belts of Ghana ........................................................ 43
    2.2    Materials and methods ......................................................................... 44
            2.2.1    Communities within regions and districts covered ................ 44
            2.2.2    Analysis of samples .............................................................. 45
    2.3    Results and discussions ........................................................................ 46
            2.3.1    Categorization of wells ......................................................... 46
            2.3.2    Occurrence of manganese and iron ........................................ 46
            2.3.3    Occurrence of arsenic. .......................................................... 48
            2.3.4    Arsenic contamination, mobilization and geology .................. 50
    2.4    Conclusions ......................................................................................... 54
    2.5    References ............................................................................................ 55

3    **Adsorptive removal of manganese (II) from the aqueous phase using
     iron oxide coated sand** ................................................................ **57**
     3.1    Introduction ........................................................................ 58
     3.2    Theoretical background ...................................................... 59
            3.2.1    Solubility and oxidation states of manganese ............. 59
            3.2.2    The adsorption phenomenon ...................................... 60
            3.2.3    Freundlich's adsorption isotherm ............................... 62
            3.2.4    Adsorption kinetic models ......................................... 62
     3.3    Materials and methods ....................................................... 65
            3.3.1    Materials for batch studies ........................................ 65
     3.4    Results and discussions ..................................................... 67
            3.4.1    The effect of pH and hydrogen carbonate .................. 67
            3.4.2    Adsorption isotherms ................................................ 68
            3.4.3    Adsorption kinetic study ........................................... 72
     3.5    Conclusion .......................................................................... 74
     3.6    References ........................................................................... 75

4    **Manganese adsorption characteristics of selected filter media for
     groundwater treatment: equilibrium and kinetics** ......................... **77**
     4.1    Introduction ........................................................................ 78
     4.2    Theoretical background ...................................................... 79
            4.2.1    Oxidation of manganese ............................................ 79
            4.2.2    Isotherms of manganese adsorption on media ............. 80
            4.2.3    Kinetics of manganese adsorption .............................. 80
     4.3    Experimental section .......................................................... 81
            4.3.1    Materials ................................................................... 81
            4.3.2    Methodology ............................................................. 82
     4.4    Results and discussion ........................................................ 83
            4.4.1    Adsorption isotherms ................................................ 83
            4.4.2    Manganese removal under oxic and anoxic condition ... 87
            4.4.3    Adsorption kinetics ................................................... 89
            4.4.4    Relationship between the manganese adsorption potential and
                     the media characteristics ........................................... 91
     4.5    Conclusion .......................................................................... 91
     4.6    References ........................................................................... 92
            Appendix 4.1 ...................................................................... 94

5    **Manganese removal from groundwater; problems in practice and
     potential solutions** .......................................................................... **95**
     5.1    Introduction ........................................................................ 96
     5.2    Theoretical background ...................................................... 98
     5.3    Materials and methods ....................................................... 99
            5.3.1    Pilot plant experiments ............................................. 99
            5.3.2    Laboratory and bench scale investigations ................ 100
            5.3.3    Development coating and Scanning Electron Microscopy
                     investigations ........................................................... 101
     5.4    Results and discussions ...................................................... 101

          5.4.1    Pilot plant experiments ................................................. 101

          5.4.2    Laboratory and bench scale investigations .................... 103

          5.4.3    Development coating and Scanning Electron Microscopy

                      investigations .................................................... 105

  5.5    Conclusions ............................................................... 109

  5.6    Acknowledgement ........................................................ 110

  5.7    References ................................................................... 110

**6    Optimising the removal of manganese in UNESCO-IHE arsenic removal family filter treating groundwater with high arsenic, manganese, ammonium and iron ................................................. 111**

  6.1    Introduction ................................................................ 113

  6.2    Theoretical background ................................................. 115

          6.2.1    The role of metallic oxides in arsenic removal .......... 115

          6.2.2    Influence of ammonium and methane on manganese,

                      arsenic and iron removal ...................................... 116

  6.3    Family filter for arsenic removal ..................................... 118

  6.4    Methodology, equipment and materials............................ 120

          6.4.1    Column experiment............................................... 121

  6.5    Result and discussions ................................................. 125

          6.5.1    Column experiments. ........................................... 125

  6.6    Conclusions ............................................................... 136

  6.7    References ................................................................... 138

          Appendix 6.1 ............................................................... 140

**7    Oxidation of adsorbed ferrous and manganese ions: kinetics and influence of process conditions........................................................... 141**

  7.1    Introduction ................................................................ 142

  7.2    Theoretical background for kinetics ................................. 143

  7.3    Material and methods ................................................... 145

          7.3.1    Preliminary tests ................................................. 145

          7.3.2    Experimental set up............................................. 146

  7.4    Results and discussions ................................................ 150

          7.4.1    Characterization of the IOCS media ........................ 150

          7.4.2    Residence time distribution.................................... 150

          7.4.3    Column experiments – adsorbed ferrous oxidation.... 151

          7.4.4    Column experiment – adsorbed $Mn^{2+}$ oxidation with

                      Aquamandix .................................................... 157

  7.5    Conclusions ............................................................... 158

  7.6    References ................................................................... 159

          Appendix 7.1 ............................................................... 160

**8    Summary and conclusions ............................................... 161**

  8.1    Introduction ................................................................ 162

          8.1.1    Groundwater quality ............................................ 162

          8.1.2    Groundwater treatment ........................................ 163

          8.1.3    Need for research and objectives ........................... 164

8.2    Presence of manganese, iron and arsenic in the groundwater.................... 165
8.3    Adsorptive removal of manganese(II) from the aqueous phase
       using IOCS ....................................................................................... 166
8.4    Manganese adsorption characteristics of selected filter media for
       groundwater treatment: equilibria and kinetics ................................... 167
8.5    Manganese removal from groundwater; problems in practice and
       potential solution .............................................................................. 168
8.6    Optimising the removal of manganese in UNESCO-IHE arsenic
       removal family filter treating groundwater with high arsenic,
       manganese, ammonium and iron ......................................................... 170
8.7    Oxidation of adsorbed ferrous and manganese ions: kinetics and
       influence of process conditions .......................................................... 172
8.8    Reference........................................................................................... 173
8.9    Recommendations for further studies .................................................. 173

Samenvatting (Summary in Dutch) .......................................................................... 175

List of symbols ...................................................................................................... 180

Abbreviations ........................................................................................................ 181

List of Publications and Presentations ..................................................................... 182

Curriculum Vitae .................................................................................................... 183

# ACKNOWLEDGEMENTS

I would like to express my gratitude to my promoter Prof. Dr. ir. J. C. Schippers and my mentor Assoc. Prof. Dr. ir. B. Petrusevski for their shepherding roles and critical comments that have helped shapened this study. There were a lot of challenges but your unflinching support, experience and encouragement has been solace to me.

To Dr. S. Sharma, I say thank you for the pieces of advice during the initial stages of the study. I am also grateful to Assoc. Prof. N. Trifunovic and Mr. J. P. Buiteman for their support, advices and seeing in me a potential to be tapped.

I gratefully acknowledge with thanks the financial support of Nuffic, the Dutch government, Kiwa Water research and the Department of Civil Engineering, KNUST, Ghana. I acknowledge with thanks the valuable support of all the staff of the WRES section of the Department of Civil Engineering, KNUST.

I also owe special thanks to three M. Sc. students namely Ms. Afiba Addae Mensah, Mr. Raktim Barua and Mrs Shrestha Rupa for their valuable contributions. To Messrs. Bruce and Parker, the laboratory staff of the Environmental and Water Quality laboratory of the Civil Engineering Department of KNUST, I say thank you. I am also grateful to Messrs. Kwame Boakye, Klutse and Nana Boadu for chauffering me around several regions in Ghana for sampling during the initial phase of the study.

My sincere thanks and appreciation to the UNESCO-IHE laboratory staff namely Drs F. Kruis, Messrs D. Galen, P. Heerings and F. Wiegman. Ms Lyzette Robbemont, thank you very much for very friendliness and support. Thanks are also due to Mr. A. Mulder of the Geoscience department of TU, Delft for your patience and assistance.

Thanks to my colleagues Ebenezer Ansa, Christina Kayoza, Sandra Irobi (TU, Delft) Kwabena Nyarko for their direct and indirect help. To my pastor and his wife (the Nwosu family) and all church members of the Mt. Zion Parish – The RCCG, I salute you all for your encouragement and prayer support.

To my father Mr. F. B. Buamah, siblings namely Theresa, George, Rosemary and Michael, I say thank you for your support and love.

My sincere thanks and appreciation go to my beautiful wife Veronica and sons – Joshua, Emmanuel, Stephen and Uriel for their love, prayer support, patience and cooperation. I love you all.

Above all, my heartfelt thanks go to God almighty for his love, support, protection and providence.

# ABSTRACT

Safe water is vital for improving the health and quality of life of people and eventually alleviates poverty. Water related diseases account for a high percentage of death in most rural and the poor peri-urban areas in the world. For example, it is an established fact that about 20% of deaths in Africa are water related. Such high incidence of water related diseases affects labour force, reduce productivity of industry and agriculture and put stress on budgetary resources. In the light of this, the presence of contaminants e.g. arsenic, manganese etc. are increasingly commanding the attention of water treatment operators and regulators. High concentrations of manganese (i.e. > 0.4 mg/l) and arsenic (i.e. > 0.01 mg/l) in drinking water are subjects of health importance. High concentration of arsenic in drinking water is known to be the cause of several incidences of various cancers in consumers over a long period of time. Chronic exposure to high manganese (>0.4 mg/l) has the potential of generating a neural disease condition similar to Parkinsonism. In addition the presence of manganese together with iron impacts bad taste to the drinking water and stains laundry and plumbing fixtures.

The provision of safe water for public consumption starts with knowledge of the quantity and quality of the source water followed by designing and development of appropriate technology to deal with the contaminants. Various assessments of water resources that have been conducted indicate that hydrological and hydrogeological data are lacking in many parts of the world.

Presently about 45% of the total drinking water production in Ghana is from groundwater sources. A perusal of the general groundwater quality data of Ghana indicates that there are groundwaters in certain parts of the country which possess potential health hazards and / or aesthetic defects due to the presence of high concentrations of manganese, arsenic, iron, fluoride, nitrate etc. High levels of arsenic exist in some of the groundwater in Obuasi, a town in the Ashanti region of Ghana (i.e. Southern part of Ghana) renowned for its gold mining activity (EAWAG, 2000). However scanty information exists regarding the presence of arsenic in other parts of the country. Apparently very limited number of investigations has been conducted on the groundwater in other parts of the country for the presence of arsenic. In the mobilisation of arsenic in groundwater, the pH of the aquifer, the redox potential, iron and manganese play important roles. Information on these parameters for the various aquifers in the country is also not adequate. This situation justifies the proposed study focused on collecting well water samples, analyze and gather additional information on the levels of manganese, arsenic, iron and other quality parameters of the groundwater used for drinking water.

Aside the health hazard that could emerge from the contaminants in drinking water, consumers' expectations of the aesthetic quality of tap water continues to increase as their quality of life improves. Consequently stringent water quality guidelines have been proposed. These demands require that water treatment plants put in upgraded / innovative technology to meet the challenging stringent requirements. Since the 1990s, the UNESCO-IHE, has been involved in research studies focusing on development of technologies for removing dissolved iron and arsenic from water. Generally, aeration followed by rapid sand filtration is commonly applied in most groundwater treatment plants for the simultaneous or successive removal of iron and manganese. Sharma (2002) studying adsorptive iron removal from groundwater showed the mechanisms involved in adsorptive iron removal and means that further improve upon the iron removal processes in practice e.g. shortening ripening time, improving filtrate quality and reducing filter backwash frequency. A similar study on manganese removal is lacking, while there is a strong need from practice to improve this process. Frequently encountered problems in practice include:

- gradual loss of manganese removal efficiency and;
- very long start-up periods when filter media has been replaced.

In addition, during extended field tests of the UNESCO-IHE – Family filters for arsenic removal from groundwater, a problematic manganese removal has been observed. The specific objectives of this research are therefore set as follows:

1. To screen the groundwater quality with highlights on the presence of manganese, arsenic, iron content in selected regions of the gold-belt zone Ghana. And to identify the geological formations associated with the contaminated aquifers.
2. To determine manganese adsorption capacities of iron oxide coated sand (IOCS) under various process conditions.
3. To determine the effect of pH on adsorption capacities of selected media for manganese and to model the adsorption phenomenon.
4. To study the rate of adsorption of Mn (II) onto one or more selected media under different oxic conditions.
5. To study the effect of manganese and iron loading on the formation of a catalytic manganese oxide coating and the subsequent effect on the start-up of manganese removal in pilot rapid sand filters.
6. To study the release of manganese from filter media and investigate measures to optimize the performance of the UNESCO-IHE family filter in the removal of manganese, arsenic and iron when treating water with high ammonium content.
7. To investigate the oxidation kinetics of adsorbed iron and manganese at different aqueous pH values.

To provide additional information on groundwater quality in the gold-belt zone of Ghana, nearly 290 well water samples from three regions namely Ashanti, Western and Brong-Ahafo, were collected and analyzed for presence of manganese, iron and arsenic. Thirteen percent of the wells in Ashanti and 29% in the Western region exceeded 0.4 mg/l – the WHO health-based guideline value for manganese. Brong-Ahafo, Ashanti and Western, regions had 5%, 25%, and 50%, of wells, respectively with iron levels above 0.3 mg/l, the Ghana drinking-water guideline value commonly accepted for iron. It was found that 5 – 12 % of all the sampled wells had arsenic levels exceeding the 10 μg/l - WHO provisional guideline value. It is estimated that between 500 000 – 800 000 inhabitants in the communities covered in this study, use untreated water with [As] > 10 μg/l. Communities within the studied area with high arsenic presence in their groundwater are located within the Birimian and Tarkwaian geological formations. Most of these arsenic contaminated wells (70%) have been in use for more than 15 years.

In studying the manganese adsorption capacity of IOCS under various dissolved oxygen conditions and pH, the IOCS demonstrated an increasing adsorption capacity ('K' values: 4.73 – 147) of Mn (II) from pH values 6 to 8. For the initial short term, comparable adsorption capacity was found at pH 6 under both oxic and anoxic conditions. This indicates that no significant quantities of adsorbed Mn (II) were oxidized at pH 6 to form extra capacity within that period. It was found also that alkalinity and pH values of 8 and higher markedly affected the solubility of Mn (II) that is governed by manganese carbonate; solubility is very limited (1 – 2 mg/L or lower) even at low alkalinity (60 ppm).

Kinetic studies using the Linear Driving force, and Potential Driving Second Order Kinetic (PDSOK) models revealed that the rate of manganese (II) adsorption onto aggregate IOCS declines after the initial phase likely due to the saturation of easily accessible adsorption capacities on grain surface and / or pH drop in the pores of the IOCS grains due to Mn (II) adsorption. The changing adsorption rate constants prevented the equilibrium concentration being predicted with the applied models.

The manganese adsorption capacities of several filter media were studied by means of batch experiments. The results indicated the following order of manganese adsorption capacities at pH 8: Aquamandix > iron oxide coated sand > Iron-ore > manganese green sand > Laterite > virgin sand. The experimental equilibrium adsorption data of most of the tested media fit well the Freundlich and Langmuir isotherm equations. Manganese adsorption capacities onto the media were significantly higher at pH 8 as compared with pH 6. The obtained results at pH 8 indicated auto-catalytic oxidation of adsorbed manganese.

Three kinetic models were applied to model the rate of adsorption / removal in batch tests. The Potential Driving Second Order Kinetic model gave the best fit, followed by the Dubinin-Kaganer-Radushkevisch model, while the Lagergren model demonstrated a rather poor fit.

To study the effect of manganese and iron loading on the formation of a catalytic manganese oxide coating and the subsequent effect on the start-up of manganese removal in rapid sand filters, pilot scale experiments were conducted using feed water having iron and ammonium and dosed with manganese. The development of the adsorptive/catalytic coating on the sand media in a pilot plant was very slow, notwithstanding the relatively high pH of 8. Low manganese concentration and more frequent backwashing resulted in a longer start-up period of the manganese removal. It can not be excluded, that nitrite has a negative effect as well. Bench scale tests conducted on the coated sand media from the pilot filters showed that the rate of adsorption/oxidation of manganese in the top layer of the filter bed is too low to explain the complete manganese removal in the pilot filters. It is likely that the adsorptive catalyst in the top layer has partly been covered with ferric hydroxide.

Different UNESCO- IHE family filters were run with i) IOCS, ii) Aquamandix layer on top of IOCS and iii) post sand filter having sand as media to investigate the manganese release phenomenon and develop measures to optimise the filters' manganese removal efficiency. In running the columns, model water containing high contents of manganese (1 mg/l), arsenic (200 µg/l), iron (5 mg/l) and ammonium (4 mg/l) were used. In the filter 1 with IOCS, after a satisfactory initial Mn removal, Mn was subsequently released into the filtrate; the As and Fe removal efficiencies were 94% - 99% for one week but dropped to (70 – 95%) afterwards. The Filter 2 equipped with a polishing layer of Aquamandix on the top of IOCS consistently removed As (95 – 99% removal) and Mn (90 – 100% removal) below the WHO standards of 0.01 mg/l and 0.4 mg/l respectively throughout the experimental period. Iron was also removed consistently (95 – 100% removal) below 0.3 mg/l, the guideline value beyond which consumers complain. However, the filtrate of Filter 2 had high amount of $NO_2^-$ (averagely 2.15 $mgNO_2^-/l$). The filter 3 comprising a post sand filter installed in series to the Filter 1 removed As, $NH_4^+$ and $NO_2$ below WHO standards. Iron was removed by the filter 3 to levels below 0.3 mg/l. Average nitrite concentrations in filtrate of the filter 3 were however high (i.e. 2.15 $mgNO_2^-/l$). The filter 3 did not remove manganese until after 7 weeks of continuous filter operation at pH 6.8. This indicates that the formation of catalytic manganese oxides at about pH 7 is slow. Manganese loading plays a role in the development of the catalytic manganese oxides however the pH may be the major determinant factor. Polishing sand layer placed on top of IOCS in the Filter 1, removed micro- flocs of Fe with attached As. The high ammonium content (4 mg/l) of the feedwater caused the fast depletion of dissolved oxygen within the columns and the establishment of anoxic conditions with the attendant high nitrite concentrations within the filtrates. The main cause of the high concentration of nitrite is likely the unavailability of sufficient oxygen for conversion of nitrite to nitrate and / survival of nitrobactor. The UNESCO-IHE family filter with addition of Aquamandix layer on the top of the IOCS showed best results by way of arsenic, manganese and iron removal when treating groundwater with high $NH_4^+$. A polishing post sand filter might be required to avoid formation of nitrite.

The oxidation of the adsorbed $Fe^{2+}$ and /or $Mn^{2+}$ on the filter media plays an important role in the removal of dissolved iron and manganese in water. The $Fe^{2+}$ in aqueous solutions is rapidly oxidized at pH value of 5 to 8. In contrast $Mn^{2+}$ is not oxidized in this pH range at all. In practice, manganese is removed in rapid sand filters at pH values higher than 6.9. This phenomenon indicates that adsorption of $Mn^{2+}$ on filter media accelerates the rate of oxidation. To investigate the rate of oxidation of the adsorbed $Fe^{2+}$ and $Mn^{2+}$ and the effect of pH on the process, short column tests were done. Iron oxide coated sand and Aquamandix were loaded with $Fe^{2+}$ and $Mn^{2+}$ under anoxic conditions. Subsequently the adsorbed $Fe^{2+}$ and $Mn^{2+}$ were oxidized by passing aerated water at pH 6 and 8. The oxygen consumed by the adsorbed ferrous on the IOCS increased about four fold upon increasing the pH from 6 to 8; 8.6% and 29.2% of the adsorbed ferrous got oxidized respectively. The real difference in the oxidation rate is most likely much higher since IOCS turned to buffer the pH during the oxidation test through dissolution of calcium carbonates and probably ferrous carbonates. Adsorbed $Mn^{2+}$ on Aquamandix did not demonstrate any uptake of oxygen within the test period. This unexpected result suggests that $Mn^{2+}$ being adsorbed onto the Aquamandix, is not a guarantee that it will be oxidized rapidly.

# CHAPTER 1

# INTRODUCTION

## 1.1    Water consumption, resources and quality

That water is very essential for human health, agriculture, industry, ensuring integrity and sustainability of the Earth's ecosystem is an undisputed fact. However this precious commodity is now running scarce in many regions of the World (WWF, 1998). And yet the availability of water is too often taken for granted. The UN environmental report states that global water shortage represents a full-scale emergency where the world water cycle seems unlikely to be able to adapt to the demands that will be made of it in the coming decades (UNEP, 1999). Water consumption has almost doubled in the last fifty years. Meanwhile, water quality continues to worsen. Even where there is enough water to meet current needs, many rivers, lakes and groundwater resources are becoming increasingly polluted. Key forms of pollution include:

- heavy metals like manganese, arsenic and iron from industry and weathered bedrock;
- fecal coliforms;
- industrial organic substances;
- acidifying substances from mining activities and atmospheric emissions;
- ammonia, nitrate, phosphate and pesticide residues pollution from agriculture;
- sediments from human-induced erosion to rivers, lakes and reservoirs;
- salinization;
- occasionally traces of radionuclides like uranium-238, radium-226, radon-222, etc.

A recent edition of the WHO's 'Safer water, better health' report has revealed that a tenth of the global burden of disease could be prevented by improvements related to drinking water, sanitation, hygiene and water resources management. Highlighting on the plight of Africa, the report indicated that 15 – 20% of deaths in Africa are water related (WHO – 'Safer water, better health', 2008). With a population of 922 million and a population growth rate of 5%, such high percentage of deaths in Africa could translate into huge absolute numbers. We cannot therefore underestimate the importance of safe water (and of course hygiene) to the health of a population.

Rapid growth of the world's population has been and will continue to be one of the main drivers of changes to patterns of water resource use. As population increases the freshwater demand increases and supplies per person inevitably decline. An alarming projection made by Gardner-Outlaw and Engelman (1997) indicates that nearly 7 billion people in sixty countries will live water-scarce lives by 2050. Even under the most optimistic projections, just under 2 billion people in forty-eight countries will struggle against water scarcity in 2050.

The twentieth century saw unprecedented economic growth. Much of this growth, was dependant on water consumption, as industries (and their demand for water) have been growing at a very fast rate. Besides the pressure exerted on water resources by increasing demand, industrialization poses a great threat to water quality. The threat of pollution of water resources comes not only from the regular operation of the industries but also from the risk of the weathering of naturally existing bedrock. As the quality of our water gets degraded and the natural ecosystems on which people and life depend get modified, our very survival gets threatened.

Some 96.5 % of the total volume of the world's water is estimated to exist in the oceans and only 2.5% as freshwater. Nearly 70% of this freshwater is considered to occur in the ice sheets and glaciers in the Antarctic, Greenland and in mountainous areas, while a little less than 30% is calculated to be stored as groundwater in the world's aquifers.

Since earliest antiquity, the human race has obtained much of its basic requirement for good-quality water from subterranean sources. Springs, the surface manisfestation of underground water, have played a fundamental role in human settlement and social development. But for many millennia, capability to abstract this vital fluid was tiny in comparison to the available resource.

Heavy exploitation followed major advances in geological knowledge, well drilling, pump technology and power development, which for most regions dates from the 1950s (Foster et al., 2000). With today's global withdrawal rate of $600 - 700km^3$ / year (Zektser and Margat - UNWWD, 2003), groundwater is the world's most extracted raw material. Many countries in the world rely on groundwater to a large extent as a source of drinking water. Presently approximately 2 billion people rely on groundwater as the only source for drinking water. Groundwater formed the cornerstone of the Asian's green agricultural revolution, provides about 70% of piped water supply in the European Union, and supports rural livelihoods across extensive areas of sub-Saharan Africa. Globally groundwater is estimated to provide about 50% of current potable water supplies, 40% of the demand of self-supplied in industry and 20% of water use in irrigated agriculture (Zektser and Margat - UNWWD, 2003).

While the quality of unpolluted groundwater is generally good, some groundwater naturally contains trace elements, dissolved from the aquifer matrix, which limit its fitness for use (Edmunds and Smedley, 1996). These elements can render water unwholesome for domestic use or may pose a public health hazard (iron, fluoride, arsenic, manganese etc.) when present at high levels. With the introduction of more systematic and comprehensive analysis of groundwater resources, supported by hydrogeochemical research; detailed knowledge of their origin and distribution is steadily increasing with the hope that associated problems can either be avoided or treated on a sound footing in the future.

Aquifers are much less vulnerable to anthropogenic pollution than surface water bodies, being naturally protected by the soil and underlying vadose (unsaturated) zone or confining strata. But, as a result of large storage and long residence times when aquifers become polluted, contamination is persistent and difficult to reverse (Clarke et al., 1996) The millenium development goal No. 7 as stipulated by the UN in the year 2000, says 'Halve, by 2015, the proportion of people without sustainable access to safe drinking water and basic sanitation'. Currently about 1 billlion of people lack access to safe drinking water; To achieve the millennium developmental goal, an additional 1.5 billion people will require access to some form of improved water supply by 2015, this is an additional 100 million people each year (or 274000/day) until 2015 (http://www.mdgmonitor.org, 2007).

Knowledge of water resources is only as good as the available data, but the various assessments of water resources that have been conducted, together with other surveys, invariably indicate that hydrological data, including hydrogeological data, are lacking in many parts of the world. Indeed, it is a twin paradox that those areas with most water resources, namely mountains, have the least data and that the nations of Africa, where the demand for water is growing fastest, have the worst capabilities for acquiring and managing water data. This lack of data applies to surface water and groundwater and to quantity and quality. There are many countries with no data on water chemistry, productivity, biodiversity, temporal changes and similar biological expressions of the state of the aquatic environment. Systems for storing, processing and managing these data and using them for assessing water resources are often rudimentary.

## 1.1.1 Groundwater use and quality in Ghana

Africa, the world's second most populous continent, with a land area of about 30.2 million $Km^2$ (11.7 million square miles) is home to about 922 million people (as of 2005) in 46 countries and 53 including all the island groups. This population forms about 14.2% of the world's human population (World population prospects: The 2006 Revision United Nations – Department of Economic and Social Affairs). The West African sub-region of which Ghana is a part, consists of sixteen countries and is home to over 230 million inhabitants. With a population growth rate of 5%, the region's population is expected to hit the 500 million mark by the year 2025 (The Hague – From vision to action, 2000). The availability and accessibility of improved water resources for all purposes will thus represent one of the main challenges to be faced by the countries.

According to the UN Statistics Division, as at 2007 the population of Ghana stood at 23.5 million with the population density estimated as 85 – 93 inh/km$^2$ and the GDP per capita was US$2660. The total surface area of Ghana is 238,533 Km$^2$. The country has total internal renewable water resources of 30.3 km$^3$/year and the groundwater internally produced stands at 26.3 km$^3$/year. Surface water produced internally stands at 29 km$^3$/year (Water resources – FAO, 2002). Presently, 70 – 89% of the population has access to improved drinking water sources (mdgmonitor.org, - website_2007).

The development of public water supplies in Ghana began in 1928. In 1965 the Ghana Water and Sewerage Co-operation (GWSC) was created by an Act of Parliament (Act 310) as a legal public utility entity charged with the responsibility of providing and managing water supply and sewerage services for domestic and industrial purposes. In July 1999 GWSC became Ghana Water Company Limited (GWCL) having been converted into a limited liability company. In Ghana, sources of water include surface water, groundwater and rain-water. Until 1994 GWCL had been operating and maintaining 208 pipe-borne and some 6500 hand – pumped borehole systems which serve an estimated 50% of the rural population throughout the country (GWCL – At a glance, 2000). Of the 208 pipe borne systems, 122 serve urban communities while the remaining 86 serve the rural communities. About 70% of the GWCL supply capacity serves the metropolitan cities of Accra, Tema, Kumasi and Sekondi - Takoradi, which constitute about 43% of the total population receiving or enjoying piped water (GWCL, 2000).

In 1994 majority of the hand – pumped borehole systems were transferred to Community Water and Sanitation Division which later developed into an agency known as Community Water and Sanitation Agency – CWSA (Act 564, 1998). The CWSA was charged with the duty of operating, maintaining and ensuring sustainability of these facilities through community ownership and management. To complement the efforts made by the GWCL, the CWSA has undertaken the drilling of more boreholes. Currently more than 12000 boreholes country-wide, especially in the rural areas, are functional and provide potable water from groundwater extraction.

Presently about 45% of the total drinking water production in the country is from groundwater sources. A perusal of the general groundwater quality data of Ghana indicates that there are ground waters in certain parts of the country which possess potential health hazards and or sensory defects due to the presence of high arsenic, fluoride, nitrate, iron and manganese levels. It has been reported that high levels of arsenic exist in some of the groundwater in Obuasi, a town in Ashanti region of Ghana (i.e. Southern part of Ghana) renowned for its gold mining activity (EAWAG, 2000). However scanty information exists regarding the presence of arsenic in other parts of the country. Apparently very limited number of investigations has been conducted on the groundwater aquifers in other parts of the country for the presence of arsenic. Studies and survey work undertaken by the Water Research Institute (CSIR) - Ghana from the 1996 up to date, have revealed five basins with high iron content in the country. The basins include: Ankobra, Tano, Ayensu, Pra, Main Volta and Lower Volta.

## 1.1.2 Groundwater use and quality in the Netherlands

The Netherlands is a small, densely populated country with an area of 37,400 $km^2$ (including inland water) and a population of 16.3 million and an average population density of 465 per $km^2$ (source: Water in the Netherlands, 2004 - 2005, 2008; Mostert, 2006). The country lies in the delta of the three major North West European rivers: the Rhine, the Meuse and the Scheldt. About 99% of the population has access to public water supply services. In the Netherlands, drinking water is obtained from groundwater or surface water; roughly two-thirds of the country obtains its drinking water from groundwater. The ground water level varies from 0.5 to 1.0 m below the soil surface in the western parts of the land; in the higher areas from 1.0 to 20.0 m.

Presently there are ten water supply companies in the Netherlands supplying 1.2 billion $m^3$ water annually (Vewin, 2008). A national groundwater quality monitoring network (NGMN) was established in the period 1979 – 1984 with the mandate of: i) describing and diagnosing the present groundwater quality in relation to land use, soil type and geohydrological conditions, ii) identifying the long term changes in equality, iii) indicating the extent of human influence on groundwater quality and providing data for evaluation of protection policies. The number of observation wells is 390 (Reijnders et al, 2007). Groundwater extracted in the Netherlands contains $0 - 30$ mgFe/l (mean 4.8 mgFe/l), $0 - 2$ mg Mn/l (mean 0.2 mgMn/l), and $0 - 35$ mgNH$_4^+$/l (mean 0.6 mgNH$_4^+$/l).

### 1.1.3 Groundwater quality problems

Qualitative problems associated with groundwater are of natural and /or anthropogenic origin. Presence of Fe and Mn for instance could confer colour, poor bitter taste, staining of laundry and plumbing fixtures (Appelo and Postma, 1994) and people will likely go back to the traditional unsafe surface water sources. Fe or Mn that enter the distribution network has a high potential of causing an increase in turbidity. The gradual accumulation of the iron deposit may reduce the carrying capacity of the distribution network and even could eventually clog them after a long period of time. Moreover iron deposited in networks during minimum flows may be re-suspended during peak hour flow thereby increasing the water turbidity. To avoid complaints by consumers, it is recommended that the Fe and Mn concentrations in potable water should not exceed 0.3 mg/l and 0.1 mg/l respectively (WHO, 2006; Same standard applied by GSB – Ghana, 1998). The WHO health based guideline value for manganese is 0.4 mg/l. Chronic exposure to high manganese (> 0.4 mg/l) over the course of years has been associated with toxicity to the nervous system producing a syndrome that resembles Parkinsonism (Fact Sheet 2001). For iron, there is currently no health based guideline of WHO. The European Union Council Directive (1998) has a more stringent guideline value of 0.2 mg/l and 0.05 mg/l for iron and manganese respectively.

Arsenic, on the other hand, if present does not pose any aesthetic problem, but may be a potential health hazard if even one takes drinking water having arsenic concentration less than 10µg/l (within the range $5 - 10$ µg/l) for a long period of time. The WHO has the value of 10µg/l as the recommended guideline value (WHO, 2006). Forty to hundred per cent of consumed, soluble forms of arsenic are readily absorbed in the human gastrointestinal tract (USEPA, 1993; Aberrantly, 1993). Long term exposure to very low arsenic concentrations causes cancer of the skin, lungs, urinary bladder, kidney and skin pigmentation. Symptoms of chronic arsenic poisoning may take between five to fifteen years to manifest physically. Doses as low as $2.0 - 4.5$ mg/Kg body weight are reported to be lethal for humans (EAWAG, 2000). WHO acceptable skin cancer risk calculated for it, is 0.17 µg/l of drinking water therefore exposure to 10 µg/l drinking water for 70 years gives a lifetime skin cancer risk of $6 \times 10^{-4}$. Recent research carried out in Thailand has indicated that chronic arsenic exposure as shown by hair samples was related to retardation of intelligence in children (Siripitayakunkit et al. 1999).

## 1.2    Manganese, arsenic and iron removal methods

In the conventional treatment of groundwater for drinking water production, aeration – and sand filtration (occasionally supported by chemical oxidation and / or sedimentation) are normally employed to remove dissolved / oxidised manganese, iron, colour and turbidity. The major advantages of this treatment scheme include its simplicity, low investment and operational cost and no or less application of chemicals (when compared with conventional treatment of surface water). A number of reports in literature have claimed that the oxidation and removal of dissolved $Fe^{2+}$ and $Mn^{2+}$ is facilitated by microbial activity (Mann et al. 1988, Mouchet, 1992, Zhang et al, 2002).

There are however several evidences and reports indicating that the removal of $Fe^{2+}$ and $Mn^{2+}$ are governed by physical and chemical processes. (Sharma et al, 1999, Petrusevski et al, 2002).

In this perception, for the removal of iron and manganese from groundwater two different process modes identified are namely:

- oxidation / flocculation mode;
- adsorptive / oxidation mode.

For the removal of iron, both modes are applied. In the oxidation / flocculation mode, $Fe^{2+}$ is oxidized with the dissolved oxygen present in the water and / or an oxidant e.g. potassium permanganate. The oxidized $Fe^{2+}$ forms the insoluble ferric hydroxides, which is removed by sedimentation and / or rapid (sand) filtration. In the filters the agglomerated ferric hydroxide particles are removed through mechanisms including straining , interception, sedimentation, Brownian diffusion, hydrodynamic retardation, surface- interaction forces and possibly biological factors. An important disadvantage of this mode is the formation of flocs, which requires the application of low filtration rates to avoid short run lengths or application of sedimention prior to the filtration (O'Connor, 1971; Sharma 1999).

The adsorptive / oxidative mode makes use of adsorption of $Fe^{2+}$ on the filter media. The adsorbed $Fe^{2+}$ is oxidized by the introduced oxygen (i.e. the dissolved oxygen) and / or an oxidant, intermittently or continuously. Adsorbed and subsequently oxidized $Fe^{2+}$ forms a coating with a high adsorption capacity that grows during prolonged operation period. In the adsorption / oxidation mode, oxidation of dissolved $Fe^{2+}$ is inevitable, however can be minimized by lowering the pH and reducing the pre-oxidation time. In ensuring these latter conditions, high filtration rates and / or low backwash frequency can be achieved. The water quality of the groundwater resource might have some influence when operating in this mode. For example the presence of $Ca^{2+}$ (up to 200 mg/l), $SiO_2$ ( $\geq$ 40 mg/l), and $PO_4^{3-}$ ($\geq$ 1 mg/l especially for new sand) may reduce the extent of adsorption of $Fe^{2+}$ onto the coating of the sand (Sharma, 2002).

In the case of manganese, the oxidation / flocculation mode is not effective when only oxygen is used as the oxidant for the oxidation of the dissolved $Mn^{2+}$ because the rate of oxidation of $Mn^{2+}$ is very low at pH values below 9. Continuous dose of an oxidant is needed to ensure oxidation of $Mn^{2+}$.

In the adsorptive / oxidation mode two ways of oxidation of the adsorbed $Mn^{2+}$ are applied namely oxidation with:

- oxygen, introduced by aeration;
- oxidants; potassium permanganate is occasionally applied (e.g. in USA).

The oxidised adsorbed $Mn^{2+}$ forms new adsorption sites. Chemical oxidation is successfully applied in practice. Many plants don't apply chemical oxidation but rather make use of the dissolved oxygen in the water to avoid the introduction of chemicals and its associated disadvantages e.g. cost and complexity of the process. However, several of the groundwater treatment plants operating are confronted with inconsistencies e.g. (very) long start up periods with new filter media, loss of manganese removal efficiency on longer term. These phenomena are supposed to be connected with the limited rate of oxidation of the adsorbed manganese or covering of the catalyst with ferric hydroxide or bacteria. The rate of adsorbed manganese oxidation is definitely much higher than the rate of oxidation of dissolved manganese oxidation in water (i.e. homogenous oxidation) however, it's not always sufficient high. Aside these two mentioned processes, other several processes are applied in practice to remove manganese, arsenic and iron. Table 1.1 gives an overview of methods applied in practices, removal mechanisms and limitations.

Some other elements like arsenic and chromium are also removed or reduced markedly especially when the media used in the filtration step has iron oxide coating. Arsenic in ground water is effectively removed in both treatment modes. An essential condition is the presence (or dosed) of sufficient iron (II). In the flocculation mode arsenic is adsorbed / co-precipitated on / with ferric hydroxide. Effective removal occurs in the adsorptive mode as well. Arsenic is adsorbed on the iron bearing coating, which is effectively renewed by adsorption of iron (II) and subsequent oxidation. Arsenate is better adsorbed than arsenite. Presence of manganese facilitates the oxidation of arsenite.

**Table 1.1** An overview of the Mn, As and Fe removal methods in practice.

| Treatment systems | Removal mechanism | Limitation |
|---|---|---|
| *Oxidation – Filtration processes*<br>Types:<br>a. Aeration - filtration<br><br><br><br>b. Chemical oxidation – precipitation & filtration | Dissolved $Fe^{2+}$ at get oxidized by dissolved oxygen / the applied chemical oxidant to ferric hydroxides which is subsequently hydrolysed. Hydrated ferric oxides agglomerate and form flocs. The flocs are removed by sedimentation and / or filtration. Dissolved $Mn^{2+}$ get oxidized by the chemical oxidant and eventually precipitate / form particulates. $As^{3+}$ oxidises to $As^{5+}$ and get adsorbed onto ferric hydroxide flocs. | $Fe^{2+}$ is easily oxidized by oxygen and oxidants. The oxidation of manganese below pH 6 will not occur, so usually oxidants are applied. Arsenic removal depends on the level and available $Fe^{2+}$. Oxidation of arsenite to arsenate is essential.<br><br>Trihalomethanes will form if chlorine is used as the oxidant. |
| c. Adsorption – oxidation with oxygen / filtration<br><br><br><br>d. Adsorption – chemical oxidation followed with precipitation – filtration. | Dissolved $Mn^{2+}$ and $Fe^{2+}$ are adsorbed onto the filter media during filtration. Preferred pH are > pH 7 and > pH 6 for $Mn^{2+}$ and $Fe^{2+}$ respectively. Adsorbed $Fe^{2+}$ get easily oxidized with aeration. Oxidation of adsorbed $Mn^{2+}$ takes probably relatively longer time with aeration.<br><br>The chemical oxidant enhances $Fe^{2+}$, $Mn^{2+}$ and $As^{3+}$ oxidation. $As^{3+}$ and $As^{5+}$ are adsorbed onto media. | Adsorption capacity of $Mn^{2+}$ reduces and rate of oxidation slows down at lower pHs (i.e. < pH 7). At pH levels below 6.9, manganese removal is problematic in practice. Efficiency of As adsorption reduces in the presence of phosphates |
| *Removal through softening with:*<br><br>a)Zeolite  or<br><br><br><br><br>b) Lime / soda ash filter bed. (Aerobic conditions recommended) | When Zeolite is applied, $Mn^{2+}$ / $Fe^{2+}$ is removed through ion exchange. $Mn^{2+}$ / $Fe^{2+}$, $Mg^{2+}$ and $Ca^{2+}$ bind to the zeolite. To rid off these cations, the filter bed is flushed with NaCl and then with clean water.<br><br>The pH during softening usually increases beyond pH 10 resulting in $CaCO_3$ precipitation. Oxidation and subsequent precipitation of $Mn^{2+}$ is facilitated. $As^{3+}$ get oxidised to $As^{5+}$ and eventually co-precipitated. | High capital cost. Requires skilled personnel.<br><br><br><br>Relatively large volume of sludge produced. Non-consistent As removal efficiency reported. pH correction needed. |

| Treatment systems | Removal mechanism | Limitation |
|---|---|---|
| *Sequestration with poly-phosphates in conjunction with $Cl_2$.*<br><br>*(This method is not a removal method per se but does prolong turbidity and colour causing reactions from reaching completion point)* | Polyphosphates could interact with $Fe^{2+}$ and $Mn^{2+}$ and eventually delay their oxidation and subsequent formation of their respective oxide or hydroxide precipitates. Silicates can interact with the $Fe^{2+}$ in a similar phenomenon. | Polyphosphates complexes are liable to break-up with time. Mn/Fe is not removed. May enhance bacteria growth. |
| *Biological iron and manganese removal* | Oxidation and removal of Mn and Fe are enhanced by microbial activity. | Ineffective in the presence of high $NH_4^+$ and $H_2S$ concentrations. |
| *Reverse osmosis and nanofiltratio)* | Dissolved $Mn^{2+}$, $Fe^{2+}$, $As^{3+}$ and $As^{5+}$ are retained. The rejection for $As^{5+}$ is higher than for $As^{3+}$. | High running and investment cost. Requires electricity. Discharge of concentrate may be a concern. |
| *In – Situ Oxidation (sub-surface removal / Vyredox method) – Normally used mainly for removal of Fe. Mn and As removal are under research.* | Adsorbed $Fe^{2+}$ is oxidized by the oxygen present in the infiltrated water. The oxidized $Fe^{2+}$ forms new sites for adsorption. The $Fe^{2+}$ present in the subsequently abstracted ground water will be removed by adsorption. | Aquifer could theoretically get clogged. |
| *Coagulation and co-precipitation (optionally supported by pre-oxidation e.g. chlorination)* | Destabilization through reduction of surface and zeta potential followed with agglomeration and precipitation of flocs. pH 6.0 – 8.5 recommended. | The arsenic removal efficiency depends on the dose of ferric coagulant. Suitable for centralized treatment. Formation of large volumes of arsenic containing sludge. |
| *(For Arsenic ions)*<br><br>*Ion exchange* | As ions are removed through ion exchange. The exchange is a function of the net surface charge. $HAsO_4^{2-}$ bind to the resin exchanger. To rid off these anions, the filter bed is regenerated with NaCl solution and then flushed with clean water. | Predominantly effective for $As^{5+}$. Efficiency is limited by competition with other anions. Monitoring needed to avoid using resin after sulfate exhaustion, else As release may occur. |

# 1.3 Origin, mobilization and chemistry of manganese, arsenic and iron

Groundwater originates from percolation of precipitation or stream flow infiltration or in some cases from direct artificial recharge. Groundwater may form springs, swamps or feed directly into surface waters when it discharges at the surface after some resident time underground. The composition of groundwater is dependent upon the constitution of the soil (humic substances, limestone, minerals etc.), quality of the infiltrating water, retention time of the water underground and the amount of pollution added through anthropogenic activities in the environment (Water quality assessment; 1997). During percolation normally, oxygen present in the water is consumed. After disappearance of oxygen, other elements or compounds acting as electron acceptors form reduced compounds together with dissolved substances like $Fe^{2+}$, $Mn^{2+}$, $NH_4^+$, $H_2S$ and $CH_4$ etc. (AWWA, 1990). Iron and arsenic may originate from sulfides which are exposed to oxygen as well. Concentration of iron and manganese in groundwater ranges from 0 to 40 mg/l and 0 to 10 mg/l respectively with the majority containing <5 mg Fe/l and <2 mg Mn/l. The concentration of arsenic in ground water ranges from less than 0.5 to 5000 µg/l.

## 1.3.1 Oxidation and reduction of manganese in groundwater

### 1.3.1.1 Manganese Chemistry

Manganese is the the 10th most abundant element in the earth's crust and second only to iron as the most common heavy metal. On the average crustal rocks contain about 0.1% Mn (Davison, 1991). Mn can display various oxidation states (11 oxidation states) ranging from -3 to +7 (US-EPA, 1994). In nature the oxidation states +2, +3 and +4 predominates. The $Mn^{2+}$ state is the most stable in solution and can form a number of different metal complexes with ligands. Mn is easily oxidized giving rise to more than 30 known Mn oxide/hydroxide minerals (Post 1999). The thermodynamic stability of the different manganese (hydro)oxides is determined by the redox conditions, pH, temperature and oxygen partial pressure within the environ. For most natural waters the pH values range from 5.0 to 8.5 and the pε values are normally +2 to +12.

From the manganese pε – pH diagram, at low pH and low redox potential, manganese occur in solution mostly as $Mn^{2+}$; at high redox potentials, and pH above 8, $MnO_2$ often predominates in natural waters (Figure 1.1). The other manganese oxides may form at lower redox potentials in basic solutions. Aside the oxides, Figure 1.1 shows that $MnCO_3$ is stable over a wide range of redox potential and pH range of 7 – 11 signifying how readily manganese may precipitate in the presence of carbonates. The $Mn^{3+}$ ion hydrolyses readily and is thermodynamically unstable in aqueous solutions but at high pH could be stabilized in the form of MnOOH. When complexed with ligands such as EDTA or citrate, $Mn^{3+}$ can attain some degree of stability. The $Mn^{4+}$ ion generally occur in the solid form as manganese oxides.

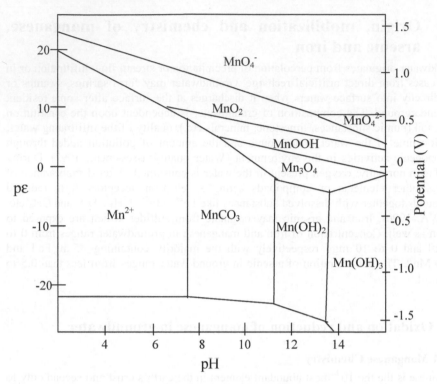

**Figure 1.1** $p\varepsilon$ – pH diagram for aqueous Mn (Stumm & Morgan, 1996).

Studies on the oxidation of Mn (II) have shown that under most laboratory conditions, manganese does not form Mn (IV), but more often the end product is one of the Mn (III) hydroxide polymorphs (Ramstedt, 2004). Recent studies suggest that microorganisms can accelerate the rate of Mn(II) oxidation by up to five orders of magnitude over abiotic oxidation and therefore are likely responsible for much natural Mn (II) oxidation (Tebo et al. 1997). In natural waters the oxidation of Mn (II) forms MnOOH instead of $MnO_2$. However at low manganese concentrations and in the presence of bacteria, amorphous Mn (IV) oxides have also been reported to form. Laboratory experiments have shown that the sorption capacity of freshly precipitated Mn oxides is extremely high for a variety of metal cations (Murray 1974, Jenne, 1968). Furthermore the adsorption of the heavy metals have been found to proceed with the release protons suggesting that the cations are bound into the Mn oxides atomic structure (Murray,1974).

### 1.3.1.2 Structure of manganese oxide minerals

Most Mn oxides occur as fine-grained, poorly crystalline aggregates and coatings. Mn oxides display a remarkable diversity of atomic structures, many of which easily accommodate a wide assortment of other metal cations. According to Post (1999) the basic building blocks for most of the Mn oxide atomic structures is the $MnO_6$ octahedron. These octahedral can be assembled by sharing edges and /or corners into a large variety of different structural arrangements, most of which fall into one of two major groups: (i) chain or tunnel structures and (ii) layer structures. There is also a third group which comprises tetrahedral and octahedral entities linked up to form a spinel-like structure (Figure 1.2).

Generally one unique characteristic of manganese oxides is the multi valence states exhibited by Mn, commonly even in a single mineral. In his study Post (1999) found that it was relatively easier to determine the average manganese oxidation but considerably difficult to determine the proportions of the Mn (IV), Mn (III) and / or Mn (II). With minerals like pyrulosite or manganosite, chemical analyses with bulk oxidation state measurements could help determine the correct Mn valence state. Other minerals may require detailed crystal structure studies to determine the valence state. X-ray spectroscopy, x-ray absorption near-edge structure spectroscopy are examples of techniques used in manganese oxidation state studies.

### 1.3.1.2.1 Tunnel Mn oxides

In the tunnel Mn oxides the $MnO_6$ octahedra building blocks are linked as single, double or triple chains sharing corners with each other to produce framework with tunnels and having square or rectangular cross sections. The larger tunnels are normally filled with water molecules and / or cations. Examples of Mn oxide in this category are pyrolusite, ramsdellite, nsutite, Hollandite group, romanechite, todorokite etc. In pyrolusite ($\beta$-$MnO_2$), single chains of edge-sharing $MnO_6$ octahedra share corners with neighbouring chains. The tunnels formed in the pyrolusite framework have square cross sections and are too narrow to accommodate other chemical species. In the case of ramsdellite the $MnO_6$ octahedral are linked as double chains to form a framework with rectangular-shaped tunnels. The tunnels normally accommodate small amounts of water, Na and Ca. In the structure of nsutite there are a blend of qualities of both the pyrolusite and the ramsdellite. The nsutite has an alternating intergrowth of ramsdellite and pyrolusite (Post 1999, Turner and Buseck, 1979).

### 1.3.1.2.2 Layer-structured Mn oxide mineral

Examples of manganese oxide in this category include lithiophorite, chalcophanite, birnessite, vernadite, buserite etc. In the lithiophorite, stack of sheets of $MnO_6$ octahedral alternate with sheets of $Al(OH)_6$ octahedral. Li cations normally fill the vacant site in the Al layer. Transition metals such as Ni, Cu and Co commonly substitute into the structure (Ostwald, 1984). The layers are cross-linked by hydrogen bonds between hydroxyl H on the Al/Li layer and O atoms in the Mn sheet. In the structure of chalcophanite edge-sharing Mn $(IV)O_6$ octahedral alternate with layers of Zn cations and water molecules. Birnessite is fine-grained and relatively poorly crystalline. Birnessite structure comprises a sheet of $MnO_6$ octahedra with interlayer cations and water molecules.

### 1.3.1.2.3 Other manganese oxide minerals

Mn oxides such as hausmanite ($Mn_3O_4$) exhibits a spinel-like structure with Mn (II) in the tetrahedral and Mn (III) in the octahedral sites. This structure has oxygen atoms in cubic close packed arrangement with Mn(III) occupying 50% of the octahedral holes and

Mn (II) in 25% of the tetrahedral holes. Other oxides like pyrochroite [Mn(OH)$_2$] consists of stacked sheets of Mn(II)(OH)$_6$ octahedra.

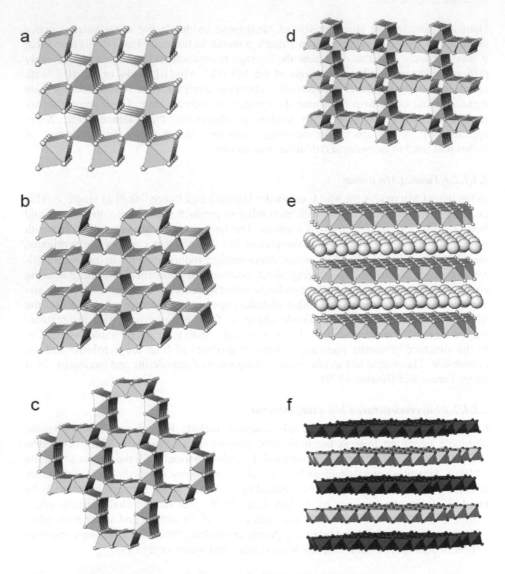

**Figure 1.2** The arrangement of MnO$_6$ octahedra in a few selected manganese oxides: a) pyrolusite; b) ramsedellite; c) hollandite; d) romanechite; e) birnessite (layers of MnO$_2$ with metal ions in the interspace; f) lithophorite layers of (Al, Li)(OH)$_2$ and MnO$_2$ alternating in the structures in c) and d). Counter ions and water have been omitted for clarity (Figure adopted from Ramstedt, 2004).

### 1.3.1.2.4 MnOOH minerals

Groutite ($\alpha$-MnOOH), Feitknechite ($\beta$-MnOOH) and Manganite ($\gamma$-MnOOH) are the three known polymorphs of the MnOOH mineral. Manganite is the most stable of the polymorphs and has high manganese and /or high sulphate concentrations. The manganite structure is similar to that of pyrulosite but all of the Mn is trivalent and one-half of the oxygen atoms are replaced by hydroxyl anions. When manganite is heated in air at 300°C it changes to pyrolusite (Dasgupta, 1965) and many crystals that appear to be manganite, but are in fact pseudomorphs, are replaced by pyrolusite. The morphological change observed from this heat induced process is a decrease in the unit cell length corresponding to the contraction of the elongated Mn(III)O$_6$ octahedral (Figure 1.2). Under acidic conditions solid manganite that reacts with protons undergoes disproportionation and goes into dissolution as Mn(II) ions and forms also one of solid Mn(IV) oxide polymorphs (Ramstedt, 2004) (Figure 1.2).

$$2 \, MnOOH(s) + 2 \, H^+ \leftrightarrow MnO_2(s) + Mn^{2+} + 2 \, H_2O \quad \log K = 7.0 - 8.6 \quad (1.1)$$

### 1.3.1.3 Manganese minerals composition and valence states

In the Table 1.2 below is a list of some of the manganese mineral with their respective states and chemical formulae. Notable among the minerals, are the Mn(IV) oxides and oxyhydroxides like pyrolusite, hausmannite, groutite, manganite that play roles in autocatalytic manganese oxidation.

**Table 1.2** Chemical composition and valence states of some manganese minerals.

| Mineral | Chemical formula | Mn Valence |
|---|---|---|
| Manganosite | MnO | +II |
| Ramsdellite | MnO$_2$ | +IV |
| Vernadite | MnO$_2$.nH$_2$O (Synthetic $\delta$-MnO$_2$ analogue) | +IV |
| Pyrolusite | $\beta$-MnO$_2$ | +IV |
| Nsutite | [Mn$^{4+}_{1-x}$Mn$_{2+x}$O$_{2-2x}$(OH)$_{2x}$] | +II, +III, +IV |
| Cryptomelane | K$_{2-x}$Mn$_8$O$_{16}$ or $\alpha$-MnO$_2$ | +III |
| Birnessite | (Na, K,Ca)$_4$Mn$_7$O$_{14}$ x 2.8H$_2$O or $\delta$-MnO$_2$ | +II, +III, +IV |
| Groutite | $\alpha$-MnOOH | +III |
| Feitknechite | $\beta$-MnOOH | +III |
| Manganite | $\gamma$-MnOOH | +III |
| Bixbyite | Mn$_2$O$_3$ | +III |
| Hausmannite | Mn$^{2+}$Mn$_2^{3+}$O$_4$ or Mn$_3$O$_4$ | +II, +III |
| Pyrochroite | Mn(OH)$_2$ | +II |
| Rhodocrosite | MnCO$_3$ | +II |
| Romanechite | Ba$_{66}$(Mn$^{4+}$, Mn$^{3+}$)$_5$O$_{10}$.1.34H$_2$O | +III, +IV |
| Todorokite | or (Ca,Na,K)x(Mn$^{4+}$, Mn$^{3+}$)$_6$O$_{12}$.3.5H$_2$O | +III, +IV |
| Hollandite | Ba$_{2-x}$Mn$_8$O$_{16}$ | +III, +IV |
| Coronadite | Pb$_{2-x}$Mn$_8$O$_{16}$ | +III, +IV |
| Manjiroite | Na$_{2-x}$Mn$_8$O$_{16}$ | +III, +IV, |
| Lithiophorite | LiAl$_2$(Mn$_2^{4+}$Mn$^{3+}$)$_6$O$_{12}$.3.5H$_2$O | +III, +IV |
| Chalcophanite | ZnMn$_3$O$_7$.3H$_2$O | +II, +IV |

## 1.3.1.4 Kinetics of manganese oxidation in aqueous solution

The essence of the oxidation of manganese is realized in tracing the origin of the metal as well as the removal processes employed at the groundwater treatment plant. The physical / chemical oxidation of dissolved manganese by oxygen in aqueous solution follows the general equation:

$$\frac{d[Mn(II)]}{dt} = - k_o[Mn(II)] + k_1[Mn(II)][MnO_x] \tag{1.2}$$

where:
$k_o = kPO_2.[OH^-]^2$
$k_o$ = reaction rate constant ($l^2/mol^2$.atm.min)
$k_1$ = reaction rate constant ($l^3/mol^3$.atm.min)
$PO_2$ = Partial pressure of oxygen (atm)

This equation implies that homogenous manganese oxidation is achieved by heterogeneous autocatalytic action, in the presence of existing Mn solid phase (Stumm and Morgan, 1996). Abiotic manganese oxidation by oxygen, in the absence of $MnO_x$ solid phase is very slow at pH values below 9 therefore manganese can not be removed quickly by simple aeration and precipitation from solutions at pHs < 8 (i.e. pH of most drinking waters) and having low alkalinity levels (i.e. below 1 mmol). To facilitate the manganese oxidation process in the aqueous phase, there is the need for a manganese solid phase eg manganese (hydro)oxides that will autocatalyze the process.

In groundwater Mn and Fe are normally supplied by upward diffusion through underlying reducing sediments. Hydrous oxides of Mn occur in most soils as discrete particles and as coatings on other mineral grains. The major Mn minerals reported in soils are lithiophorite, hollandite and birnessite (Golden et al, 1993). Birnessite have been found to directly oxidize Se(IV) to Se(VI), Cr(III) to Cr(VI), As(III) to As(V) (Bajpai and Chaudhuri, 1999; Scott and Morgan, 1996; Manceau and Charlet, 1992). At pH 7 manganese oxides are considerably more oxidizing than iron oxides and can catalyze the oxidation of As(III), Co(II), Cr(III), Pu(III) and oxidatively degrade natural humic and fulvic acids (Huang 1991, Stone and Ulrich 1989; Sunda and Kieber 1994). The highly charged surfaces of the manganese oxides make them good scavengers for a variety of trace elements like Cu, Co, Cd, Zn, certain radionuclides e.g. $^{60}Co$ and isotopes such as Ra, Th. Consequently the manganese oxides play effective role in the mobilization of these trace elements in the environment

## 1.3.1.5 Microbial oxidation of manganese

Apart from the stalked bacteria of Gallionella genus, most of the microorganisms which mediate the oxidation of Fe(II) have been found to mediate manganese oxidation. The list of microorganisms known to mediate manganese oxidation include Leptothrix, Crenothrix, Hyphomicrobium, Siderocapsa, Metallogenium, Marine Bacillus strain SGI Siderocapsa, Caulococcus etc (Mann et al. 1988, Mouchet, 1992, Dong et al. 2005). There are a number of strains of bacteria that can increase Mn oxidation rates by as much as five orders of magnitude, the rate being dependent on which species of bacteria is involved (Nealson et al. 1988; Tebo 1991; Wehrli et al 1995, Zhang et al. 2002). The degree of influence that bacteria have on Mn oxidation has proved difficult to determine, as the majority of metabolic inhibitors that are used to prevent biotic activity in control experiments also influence Mn oxidation rates (Shiller, 2004). Most of the known manganese oxidizing bacteria are aerobes and therefore applying the bateria to remove manganese from groundwater without aeration treatment can be a problem (Dong et al, 2005).

*Mechanism of microbial manganese oxidation*

Nealson et al, 1989 have reported Mn(II) oxidation in fresh – and marine water, soils and sediments by bacteria and fungi through direct and indirect extracellular mechanisms. In direct Mn(II) oxidation, the manganese get bound to cell macromolecules e.g. cell wall components, proteins etc, with simultaneous occurrence of the oxidation process. With the indirect Mn(II) oxidation the microorganism involved modifies the existing redox conditions of the local aqueous environment through release of oxidants or acids or bases (Ghiorse 1984, Nealson et al 1989). Several suggestions have been put forward to further explain the mechanism for the microbial Mn(II) oxidation. One speculation is that microorganisms first bind and enzymatically oxidize Mn(II) to Mn(III). The Mn(III) formed is in turn oxidized to Mn(IV) which precipitates as $MnO_2$ (Ehrlich 1996a,b).

$$2 Mn^{2+} + \frac{1}{2} O_2 + 2 H^+ \rightarrow 2 Mn^{3+} + H_2O \qquad (1.3)$$

$$2 Mn^{3+} + 3 H_2O + \frac{1}{2} O_2 \rightarrow 2 MnO_2 + 6 H^+ \qquad (1.4)$$

Calculating the $\Delta G^{\circ}\,'$ for reaction (1.3) at pH 7 using a $Mn^{3+}/Mn^{2+}$ redox couple, the above set of reactions appears to be highly unfavourable unless it's coupled to another reaction that can input some energy.

Another speculation that seems more feasible from the thermodynamic point of view involves a two step mechanism whereby Mn(II) is oxidized by a hydrogen peroxide to Mn(III). The Mn(III) formed remains bound to the enzyme till the subsequent oxidation of the Mn(III) to $MnO_2$ occurs.

$$2 Mn^{2+} + H_2O_2 + 2 H^+ \rightarrow 2 Mn^{3+} + 2H_2O \qquad (1.5)$$

$$2 Mn^{3+} + 4H_2O + O_2 \rightarrow 2 MnO_2 + H_2O_2 + 6 H^+ \qquad (1.6)$$

A number of alternative benefits that microbes may receive from the manganese oxidation activities have been proposed. The Mn(II) oxidation may be a mechanism by which the microbial cell protect itself from toxic oxygen species or convert soluble toxic Mn(II) into extracellular particulate MnOx. Another possible benefit may be that the metal encrusted surfaces of Mn(II)-oxidzing bacteria serve as protective cover against predation, viral attack or ultraviolet radiation (Emerson, 1989). Another proposal put forward is that the oxidation of Mn(II) may be a mechanism whereby bacteria indirectly utilize pools of carbon contained in humic substances (Sunda and Kieber, 1994).

## 1.3.2 Arsenic in groundwater

Arsenic is the 20th most abundant element in the earth's crust and the 12th most abundant element in the human body (Borgono et al. 1977; Abedin et al. 2002); it has an average crustal concentration of 1.8 mg/Kg (Demayo, 1985). However through geogenic processing of crustal materials, arsenic can be concentrated in soils to a typical range of 2 to 20 mg/Kg. Concentrations as high as 70 mg/kg have been reported by Yan-chu, (1994). Arsenic may occur as semi-metallic element ($As^0$), arsenate ($As^{5+}$), arsenite ($As^{3+}$) or arsine ($As^{3-}$). The toxicity of $As^{3+}$ is 10 times higher than that of $As^{5+}$ (Castro de Esparza, 2006). Arsenic is labile and readily changes oxidation or chemical form through chemical or biological reactions that are common in the environment. The mobility of arsenic is usually controlled by redox conditions, pH, biological activity and adsorption / de-sorption reactions. In aqueous environment arsenic normally exist in the +III or +V states. The molecular structures of different arsenic compounds are given in Figure 1.3.

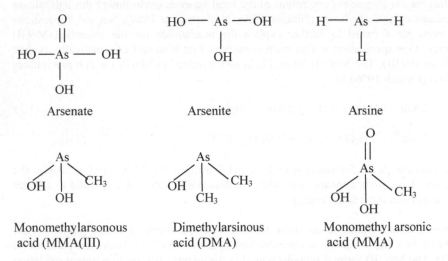

**Figure 1.3** Molecular structures of some arsenic compounds.

Arsenic in groundwater most often originates from geogenic sources, although anthropogenic pollution sources do occur (Chakravarty et al. 2003). Arsenic occurs as a major constituent in more than 200 minerals, including elemental arsenic, arsenides sulphides, oxides, arsenates, arsenites etc. Arsenic can exist in the inorganic and organic forms. The concentration of arsenic in ground water ranges from less than 0.5 to 5000 µg/l. In groundwater, the dominant As is a function of pH and redox potential. $As^{5+}$ predominates in oxic water (e.g. surface water) and depending upon the pH value, can exist as $H_3AsO_4$, $H_2AsO_4^-$, $HAsO_4^{2-}$ and / or $AsO_4^{3-}$.

In the pH range 6 – 9, $HAsO_4^{2-}$ and $H_2AsO_4^-$ dominate with relatively low concentration of $AsO_4^{3-}$ (Figure 1.4). Below pH 6, $H_2AsO_4^-$ and $H_3AsO_4$ dominate whiles above pH 9 only $AsO_4^{3-}$ mostly occur. As (III) is the dominant species under reducing conditions, and therefore it is the principal form of As in groundwater. Depending upon the groundwater pH, the arsenic can exist as $H_3AsO_3$ and / or $H_2AsO_3^-$. Below pH 9, $H_3AsO_3$ is the dominant species whereas $H_2AsO_3^-$ and $HAsO_3^{2-}$ dominate above pH 9 (Ferguson and Gavis, 1972; Sracek et al 2001) (Figure 1.4 and 1.5). It is however noted that both the As (V) and As (III) forms occur in both oxic and reducing conditions due to the slow oxidation and reduction kinetics (Edwards, 1994).

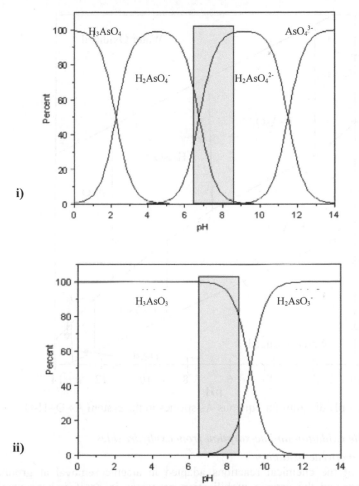

**Figure 1.4** Concentrations of the i) As(V) ii) As(III) at different pH values. The shaded area is the pH range of most groundwaters. (Wilson et al. 2003)

### 1.3.2.1 Mobilization of As in groundwaters

Several theories have been put forward to explain how arsenic gets mobilized into groundwaters. High arsenic concentration in groundwater most commonly results from:

- pyrite oxidation and de-sorption from oxyhydroxides;
- water – rock reaction with arsenic bearing minerals;
- volcanic activity;
- up-flow of geothermal water;
- anthropogenic sources.

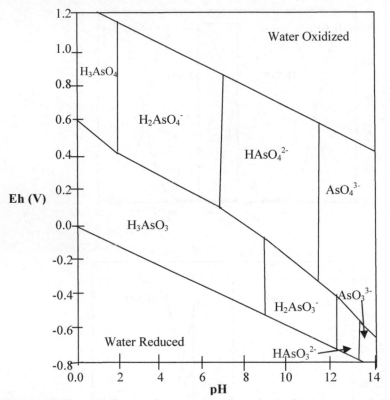

**Figure 1.5** Eh – pH diagram for aqueous As species in the system As-$O_2$-$H_2O$.

### 1.3.2.1.1. Pyrite oxidation and de-sorption from oxyhydroxides

*Pyrite and arsenopyrite oxidation*

An overview of the chemical reactions adapted in arsenic removal at groundwater treatment plants and the arsenic mobilization processes in aquifers have shown the sorption and de-sorption of arsenic onto ferric oxyhydroxides as dominant phenomena. Arsenic mobilisation in groundwater has been explained by a number of researchers to be due to the oxidation of arsenopyrite. It is believed that arsenic occur in some sediments together with pyrite and get released due to heavy withdrawal of groundwater (Acharyya et al. 1999; Das et al. 1995). The withdrawal is believed to cause lowering of the groundwater table and subsequently results in the penetration of oxygen into the deeper sediments. Alternatively removal of the top soil layer could expose (arseno) pyrite to oxygen. Eventually, the pyrites rich in arsenic get oxidized as shown in the following equation:

$$2\ FeS_2 + 7O_2 + 2H_2O \rightarrow 2\ Fe^{2+} + 4\ SO_4^{2-} + 4\ H^+ \tag{1.7}$$

$$4\ FeAsS + 11O_2 + 6\ H_2O \rightarrow 4\ Fe^{2+} + 4\ H_3AsO_3 + 4\ SO_4^{2-} \tag{1.8}$$

Aside the oxidation with oxygen, the arsenopyrite can be oxidized by $Fe^{3+}$ as follows (Sracek et al. 2001):

$$FeAsS + 13\ Fe^{3+} + 8H_2O \rightarrow 14\ Fe^{2+} + SO_4^{2-} + 13H^+ + H_3AsO_4(aq) \tag{1.9}$$

The $Fe^{2+}$ (in equations 1.8 and 1.9) formed get oxidized to $Fe^{3+}$ which then sorbs arsenic molecules released from the pyrite oxidation. Redox reactions involving pyrite, or iron oxyhydroxides derived from pyrite oxidation, are important sources and sinks of trace metals such as arsenic in sedimentary aquifers (Kolker, 2000). Studies on the groundwater in Bangladesh suggest that areas of arsenic-contaminated groundwater tend to occur in places with sediments containing relatively high concentrations of extractable Fe (and associated As). As (V) sorption by iron oxides is characterized by a highly nonlinear sorption isotherm, meaning that at very low arsenic concentrations in solution the loading of arsenic on the surface of iron oxides can be relatively high (Manning and Goldberg, 1996; Hiemstra and van Riemsdijk, 1996).

Just like Fe (III) oxides, manganese oxides may also play an effective role in the sorption of As whenever present. At neutral pH, redox reactions of arsenate (III) with manganese (IV) oxides and manganese (III) – oxides can be written as:

$$H_3AsO_3 + MnO_2 \leftrightarrow HAsO_4^{2-} + Mn^{2+} + H_2O \tag{1.10}$$

$$H_3AsO_3 + 2MnOOH + 2H^+ \leftrightarrow HAsO_4^{2-} + 2Mn^{2+} + 3H_2O \tag{1.11}$$

$\delta MnO_2$, which resembles the naturally occurring mineral birnessite, leads to a faster oxidation of arsenate (III) compared to $\alpha$- and $\beta$-modification of $MnO_2$. This is attributed to the low crystallinity and layered structure of the $\delta$-modification, which has easily available reaction sites with $Mn^{4+}$ and $Mn^{3+}$ in the interlayer of the solid (Oscarson et al 1983). From the above reactions it is expected that such oxidative process would result in the release of divalent manganese ions, however the contrary is realized in practice. Oscarson et al (1983), Moore et al (1990) and Driehaus et al (1995) all found an apparent substantial lack of release of $Mn^{2+}$ ions during oxidation of As (III), an evidence indicative of the adsorption of the As (III) to manganese oxide or the formation of an As (V)-Mn (II)-complex. Manganese compounds could also act as catalyst for the oxidation of As (III) to As (V). Curves based on Freundlich isotherms worked on by Driehaus et al (1995) confirmed that manganese oxides adsorb arsenate very quickly. Although $\delta MnO_2$ has a negative surface charge at pH > 4, due to its $pH_{pzc}$ (i.e. pH of point of zero charge) <3.5, a significant adsorption of arsenate occurs.

Adsorption of arsenate onto hydrous aluminium oxides may also be important if these oxides are available. It must be mentioned that arsenate loadings onto Al and Mn oxides are much smaller on weight basis compared to that of Ferric oxides. Adsorption reactions are responsible for the relatively low concentrations of arsenic found in most natural waters.

*Arsenic de-sorption from oxyhydroxides*
Chemical processes that cause changes in iron redox chemistry are particularly important in releasing large quantities of arsenic into surrounding waters (BGS and DPHE, 2001). A number of such processes have been elaborated as follows.

1.          Reductive dissolution of hydrous ferric oxides – The presence of fresh organic matter from buried soils and high ambient temperature (about 28 °C) tend to facilitate the microbial consumption of oxygen during process of organic matter oxidation. During this process nitrates and sulphates are reduced. Nitrates get reduced to nitrite and subsequently to dinitrogen oxides and nitrogen gas. Sulphates present get reduced to sulphides which then react with available Fe to produce FeS and ultimately pyrites. There are also produced alongside, other compounds like methane from fermentation and methanogenesis and ammonium from conversion of nitrogen. When the sediments become reduced, a series of geochemical reactions could occur that lead to release of arsenic into the aquifer. The exact processes are however not yet fully understood. The likely reactions could be any one of the following:

  (i)    reduction of strongly adsorbed As(V) to maybe less strongly bound As(III), leading to release of arsenic (Nickson et al, 1998);
  (ii)   when oxygen decreases and get to minimal levels the oxhydroxides become the electron acceptor according to the following reaction.

$$4 \text{ Fe(OH)}_3 + \text{CH}_2\text{O} + 7 \text{ H+} \leftrightarrow 4 \text{ Fe}^{2+} + 10 \text{ H}_2\text{O} + \text{HCO}_3^- \qquad (1.12)$$

  The iron(III) oxides partially dissolve and release iron(II) as well as coprecipitated and adsorbed arsenic and phosphate. Insoluble manganic oxides /oxyhydroxides (if present) dissolve by reduction and release their sorbed arsenic prior to the Fe oxyhydroxides reduction. As(V) reduction would normally occur.
  (iii)  the iron(III) oxides undergo slow geochemical changes leading to the desorption of adsorbed arsenic. (BGS & DPHE, 2001)

2.          Carbonate complex formation - Recent studies and data suggest that carbonate ions form stable complexes with As(III) and As(V) in solution. This will lower the activity of free arsenic ions in solution and will tend to increase the solubility of any As minerals and cause the desorption of As from the oxides. Bicarbonate concentrations in many affected groundwaters are relatively high as result of oxidation of fresh organic matter and the dissolution of carbonate minerals (Smedley et al. 2001).

3.          pH changes and competitive sorption - Changes in pH can lead to desorption of As(V). As(III) sorption is little affected by pH changes within the pH range of most

groundwaters. The presence of competing anions like phosphates, silicate and bicarbonate could lead to desorption of As(III) and As(V) through competitive sorption and changes in pH.

### 1.3.2.1.2 Water – rock reaction with arsenic bearing minerals

Arsenic occurs as a major constituent in more than 200 minerals including elemental arsenic, realgar, orpiment, arsenopyrite, arsenolite, scorodite etc. Most of the materials are ore minerals. The most abundant ore minerals - arsenopyrite (FeAsS), arsenian pyrite $(Fe(S,As)_2)$ and other arsenic-sulphide minerals that are normally formed at high temperatures within the earth crust are often the most important sources of arsenic in ore zones (Nordstrom, 2002). Arsenic is found at times in the crystal structure of many sulphide minerals (e.g. pyrite – FeS) as a substitute for sulphur. Pyrite commonly forms in zones of intense reduction such as around buried plant root or other nuclei of decomposing organic matter. During its formation some of the soluble arsenic may be incorporated. In oxidising environment pyrite gets oxidised to Fe oxides with the subsequent release of large amounts of $SO_4$, acidity and associated trace constituents of arsenic. The presence of pyrite as a minor constituent in sulfide-rich coals is ultimately responsible for the production of 'acid rain' and for the presence of arsenic problems around coal mines and areas of intensive coal burning.

### 1.3.2.1.3 Volcanic activity

Natural arsenic in groundwater and surface waters in the Latin Americas has been associated with volcanic activity in the Andes mountains. This process that is still continuing is shown in lava flows, fumaroles, thermal springs and geothermal phenomena. The volcanic activity may also influence certain water properties like pH. variable alkalinity, low hardness, moderate salinity and presence of boron, fluorine, silica and vanadium (Wilson et al. 2003).

### 1.3.2.1.4 Up-flow of geothermal water

High concentrations of As (up to about 370 $\mu l^{-1}$) has been reported by a number of researchers to occur in some areas as a result of inputs from geothermal sources with arsenic rich groundwaters. Arsenic associated with geothermal waters has been reported in many parts of the USA, Japan, New Zealand, Chile, Iceland, France etc.(Welch et al 1988; Criaud and Fouillac 1989; Nimick et al., 1998; Wilkie and Hering, 1998). Geothermal heat at the surface is highly concentrated at places where magma is close to the surface. This primarily occurs in volcanic and hotspot areas. Much of this heat is conducted and convected to water bodies within the earth giving rise to circulation of hot waters, hot springs etc. Some river waters affected by geothermal activity show distinct seasonal variations in their arsenic concentrations (Smedley and Kinniburgh 2002).

### 1.3.2.1.5 Anthropogenic sources

Elevated values of arsenic (up to about 30$\mu l^{-1}$) in water bodies may also occur as a result of pollution from industrial or sewage effluents (Andreae and Andreae, 1989). Arsenic can also be derived from mine wastes and mill tailings. Azcue et al.(1994) reported arsenic concentration of 556 $\mu l^{-1}$ in streams adjacent to tailings in British Columbia.

Though often involving notable increases in arsenic content of the water body involved above the baseline concentration, such anomalies tend to be relatively localised around the pollution source. Normally the localization results from the strong affinity of the oxide minerals especially Fe oxide for arsenic under oxidising, neutral to mildly acidic conditions.

### 1.3.3 Iron in groundwater: reduction and oxidation

Iron is the fourth most abundant element in the earth's crust. The major iron ores are hematite, $Fe_2O_3$, magnetite, $Fe_3O_4$, limonite, $FeO(OH)$ and siderite, $FeCO_3$. Iron is a common constituent of anoxic groundwater. In the normal pH range of groundwater (pH = 6 – 8.5), dissolved iron is present as $Fe^{2+}$. The main sources of $Fe^{2+}$ in groundwater are the dissolution of iron (II) - bearing minerals and the reduction of iron – oxyhydroxides (FeOOH) present in the sediments. Important iron (II)– bearing minerals commonly present in aquifer materials comprise minerals like magnetite, ilmenite, pyrite, siderite, and iron (II)-bearing silicates like amphiboles, pyroxenes, olivine, biotite, glauconite and clay minerals such as smectites (Appelo and Postma, 1992).

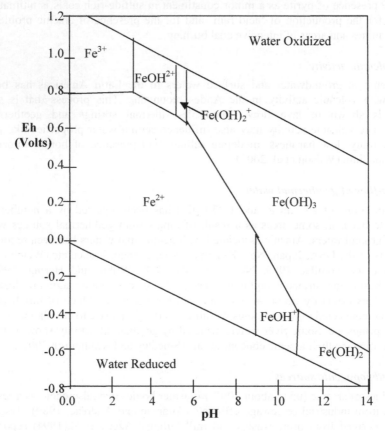

**Figure 1.6** Eh – pH diagram for aqueous Fe.

In the case of the iron (II) – bearing silicates and magnetite under oxic conditions the $Fe^{2+}$ present is oxidised and the mineral surface becomes covered by an inner layer of Fe (III) silicate and an outer layer of FeOOH which inhibits the iron dissolution process by diffusion control (Schott and Berner, 1983; White, 1990; Hochella and White, 1990). Reductive dissolution of FeOOH becomes important when an electron acceptor, such as dissolved organic matter, hydrogen sulphide or methane enters a sediment containing FeOOH. The pool of FeOOH in the sediment covers a broad range of minerals with both variable stability and variable kinetic properties. Common FeOOH minerals found in aquifers are ferrihydrite (5 $Fe_2O_3$. $9H_2O$), goethite ($\alpha$- FeOOH), lepidocrocite ($\gamma$- FeOOH) and hematite (Appelo and Postma, 1994).

Sung and Morgan (1980) reported that the initial product of iron (II) oxidation are lepidocrocite and amorphous FeOOH. These forms are later transformed into goethite by aging. Lepidocrocite is an effective catalyst for iron oxidation but goethite is not (Takai, 1973). These mineral forms may have different specific surface, surface site densities and type of sites, hence different adsorption capacities for iron (II). Stenkamp and Benjamin (1994) have reported that surface characteristics of underlying sand could also have some effect on the properties of subsequent developed surface of coated sand, if the coating is thin or porous.

In natural waters containing greater than 1mM carbonate alkalinity, ferrous carbonate complexes i.e. $Fe(CO_3)$, $Fe(CO_3)_2^{2-}$, and $Fe(CO_3)(OH)^-$, dominate the speciation of Fe(II) (King, 1998) (Figure 1.6). At pH values above 6.0, the complex - $Fe(CO_3)_2^{2-}$ is the most kinetically active specie. For pH values below 6.0, the oxidation rate of Fe (II) is well described in terms of $Fe^{2+}$ and $FeOH^+$ species.

### 1.3.3.1 Kinetics of iron (II) oxidation

The oxidation of Fe(II) by oxygen is given by:

$$4Fe^{2+} + O_2 + 2H_2O \rightarrow 4Fe^{3+} + 4OH^- \tag{1.13}$$

$$4Fe^{3+} + 4OH^- + 8H_2O \rightarrow 4Fe(OH)_3 + 8H^+ \tag{1.14}$$

$$4Fe^{2+} + O_2(g) + 10H_2O \rightarrow 4Fe(OH)_3(s) + 8H^+ \tag{1.15}$$

At a pH $\geq$ 5, the following rate law is applicable (Stumm and Lee, 1961):

$$\frac{d[Fe^{2+}]}{dt} = -k(PO_2)[OH^-]^2[Fe^{2+}] \tag{1.16}$$

where:
$d[Fe^{2+}]/dt$ = rate of Fe(II) oxidation, mol/(L)(min)
k (reaction rate constant) = 8.0 ($\pm$ 2.5) x $10^{13}$ $L^2$ $(min)^{-1}(atm)^{-1}(mol)^{-2}$ at 20.5° C
$PO_2$ = partial pressure of oxygen, atm
$OH^-$] = hydroxide ion concentration, mol/l
$[Fe^{2+}]$ = iron (II) concentration, mol/l

Thus the oxidation may be represented as first order with respect to iron (II) and molecular oxygen and second-order with respect to the hydroxide ion.

From a practical viewpoint, however, the oxidation of iron (II) may be considered to follow pseudo first-order kinetic (AWWA; 1990), as follows:

$$\frac{d[Fe^{2+}]}{dt} = - \frac{1.68 \times 10^{-15}}{[H^+]^2} [Fe^{2+}] = -K[Fe^{2+}] \qquad (1.17)$$

where K = the apparent first order reaction rate constant (i.e. $1.68 \times 10^{-15} / [H^+]^2$ min$^{-1}$); assuming that the partial pressure of oxygen is 0.21 atm and that the ionisation constant for water is $10^{-14}$.

A more recent studies by Wehrli (1990) and Millero (1990,1992) based on both free energy calculations and correlation of observed increases in iron oxidation rate to the change in iron speciation with increased pH revealed that the overall Fe(II) oxidation rate ($k_{app}$) can be explained in terms of the weighted sum of the oxidation rates of individual Fe(II) species:

$$- \frac{d[Fe^{2+}]}{dt} = k_{app}[PO_2][Fe^{2+}] \qquad (1.18)$$

where

$$k_{app} = 4(k_1\alpha_{Fe}^{2+} + k_2\alpha_{FeOH}^{+} + k_3\alpha_{Fe(OH)2} + \ldots\ldots k_n\alpha_n)$$

Equations 1.16 and 1.17 reveal that the rate of oxidation of iron(II) with molecular oxygen proceeds at an increasing rate as pH rises, other factors being held constant. The reaction may be quite slow at pH 6 and very rapid at pH 7.5 and above. Apart from pH, other factors like alkalinity (Ghosh et al 1996), temperature (Stumm and Lee, 1961: Sung and Morgan, 1980), organic matter (Theis and Singer, 1974), silica, copper, manganese, cobalt (Stumm and Lee 1961) have been reported to have significant effect on the rate of oxidation of Fe(II). Oxidation of Fe(II) has also been reported to be accelerated by the presence of Fe(III) (Tamura et al 1976, Tufekci and Sarikaya, 1996); with the reaction proceeding along two pathways:

- the homogenous reaction pathway that takes place in solution;
- the heterogenous reaction that occurs on the surface of ferric hydroxide precipitates.

At constant pH and oxygen concentration the rate is expressed by:

$$- \frac{d[Fe(II)]}{dt} = (k^\theta + k^1[Fe(III)]) [Fe(II)] \qquad (1.19)$$

The rate constant $k^\theta$ for homogenous reaction is equal to $K_o[O_2][OH^-]^2$ and $k^1$ for the heterogeneous reaction is determined by $k_{so}[O_2]K_e / [H^+]$; $K_o$ and $k_{so}$ being real rate constants and $K_e$ being the equilibrium constant for the adsorption of iron (II) onto iron

(III) hydroxides. At around the neutral pH, much of the iron (III) is in the form of hydroxide precipitate with a positive surface charge (King, 1998).

# 1.4    Theoretical background of adsorptive processes

The various adsorption principles highlighted in this section apply to the origin, mobilization and treatment processes for iron, manganese and arsenic in water.

## 1.4.1 Equilibra

Adsorption phenomena involve separation of a substance from one phase accompanied by its accumulation or concentration at the surface of another. The adsorbing phase is the adsorbent, and the material concentrated or adsorbed at the surface of that phase is the adsorbate (Slejko, 1985). The character of the quantitative equilibrium distribution between phases that markedly affects the feasibility of adsorption as a separation process for a particular application is influenced by a variety of factors. These factors relate to the properties of the absorbate, the adsorbent and the system in which adsorption occurs.

## 1.4.2 Forces and energetics of adsorption

Adsorption at a surface or interface is largely the result of binding forces between the individual atoms, ions or molecules of an adsorbate and the surface, all of these forces having their origin in electromagnetic interactions (Weber, 1972; Van Vliet and Weber, 1981). Exploitation of adsorption requires the full understanding of the electrostatic forces and the energy factor involved, their sources and the kinetics of the uptake. Time dependence of solute concentration distribution in both bulk and at adsorbent surface is also a significant factor to be considered. Interaction between adsorbate and adsorbent consist of molecular forces embracing permanent dipole, induced dipole and quadruple electrostatic effects. The success of adsorption processes in removing solutes from water depends on the magnitude of the forces at play and the adsorbent solid-solute contact time. These forces that influence the net energy of interaction of a solute with an adsorbent may be classified into two:

- The short range forces otherwise called chemical forces. These forces often give rise to covalent or hydrophobic bonds or hydrogen bonding or steric effect that manifest itself or themselves in ion exchange and chemical adsorption.
- Long range forces or the coulombic forces that lead to electrostatic and van der Waals attraction (Stumm and Morgan, 1996). Such forces facilitate physical adsorption.

Based upon these adhesive forces existing between the solute and the media grains four principal types of adsorption have been identified, these are ion exchange, physical, chemical and specific adsorption (Yang,1999). Specific adsorption not mentioned above, exhibit qualities characteristic of both short range and long range forces. Clearly, when an adsorbate molecule approaches a solid adsorbent surface, the molecule interacts with a large assemblage of atoms in the crystal lattice of the adsorbent simultaneously. The potential energy function, say Ur, (which is the sum of all interactions between an adsorbate molecule and molecules in the lattice of the adsorbent) passes through a minimum known as the potential well, the depth $Ur_o$, which is the energy of adsorption at a temperature of absolute zero (Crittenden and Thomas, 1998). The depth corresponds to several kilojoules per mole. For a given adsorbate-adsorbent system, $Ur_o$ equates closely with measured heat of adsorptions (Figure 1.7). Such heats of adsorption can be measured from calorimetric experiments or adsorption isotherms and isobars.

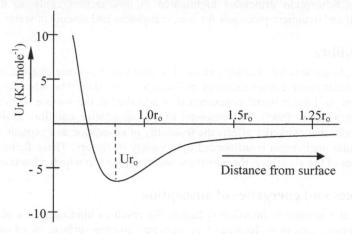

**Figure 1.7** Potential Energy diagram for adsorption.
(Ur represents potential energy; $r_o$ is the distance between surfaces of an adsorbate molecule and a molecule in the lattice of an adsorbent.)

For two atoms separated by a distance, r, the total potential energy ε, is the sum:

$$\varepsilon (Ur) = -C.\ r^{-6} + B.\ r^{-12} \qquad (2.19)$$

where C (the dispersion) is a theoretically calculated constant associated with dipole-dipole interaction and B is an empirical constant. The term $-C.r^{-6}$ is a theoretical expression for long range attractive potential. An exponential function of r, derived to express the short-range repulsive potential is approximated by the term $Br^{-12}$ in the equation above.

## 1.4.3 Adsorbate - solvent properties

The extent of adsorption relates to certain properties of the adsorbate relative to the solution phase, notably those of surface tension and solubility (Weber, 1972). For example, many organic compounds can effectively lower the surface tension of water; and the energy balance of the aqueous systems of such compounds favours their partitioning to solid-water interface. The extent of adsorption is generally greatly influenced by the lyophobicity of a compound, that is, its insolubility in the solvent phase. Bonding between a substance and the solvent in which it is dissolved must be broken before adsorption from the solvent can occur. The primary driving force for adsorption from solution therefore may relate to the lyophobic character of the adsorbate or to a particular affinity of the adsorbate for the surface of the adsorbent.

A number of parameters specific to a given system will therefore affect adsorption. They include (for the adsorbate): concentration, molecular weight, molecular size, molecular structure, molecular polarity, steric form or configuration and the nature of background or competitive adsorbates (Mattson and Mark, 1971; Weber, 1972; Van Vliet and Weber, 1981).

For the adsorbent, the most important determinants of equilibrium capacity and the rate of approach to this capacity include surface area, the physicochemical nature of the surface, the availability of that surface to adsorbate molecules or ions, and the physical sizes and form of the adsorbent particles. System parameters such as temperature and pH can markedly influence the adsorption. Molecular size can also affect adsorption rates if these rates are controlled by intra-particle diffusive mass transport within porous adsorbents (Slejko, 1985). Particles with small molecular size are generally adsorbed faster than those of bigger molecular size except for situations where a large molecular size particle possesses a higher driving force or energy for adsorption (Weber and Morris, 1963).

### 1.4.4 Adsorption isotherms

Adsorption reactions are normally exothermic, with changes in enthalpy of the order of condensation or crystallization reactions. The adsorption capacity in a given system is thus generally found to increase with decreasing temperature. Conversely, because adsorption kinetics are generally controlled by diffusive mass transfer, rates of approach to equilibrium normally increase with increasing temperature (Slejko, 1985). The adsorption of a substance from one phase to the surface of another in a specific system leads to a thermodynamically defined distribution of that substance between the phases when the system reaches equilibrium that is, when no further net adsorption occurs. The common manner in which to depict this distribution is to express the amount of substance adsorbed per unit weight of adsorbent, q, as a function of the residual equilibrium concentration, C, (or steady state equilibrium pressure) of substance remaining in the 'solution' phase. An expression of this type, termed an adsorption isotherm, defines the functional equilibrium distribution of adsorption with concentration of adsorbate in solution at constant temperature (Slejko, 1985). The adsorption capacities of most filter media are commonly described with the Freundlich's or Langmuir equations. Using these equations the adsorption capacities of various filter media under different process conditions can be determined.

## 1.5 Need for research

In Ghana, there is limited information on the presence of arsenic in the about 20 000 boreholes used for drinking water, most of them without any treatment. At the same time there are indications that in some parts of the country symptoms like arsenicosis have been identified. In addition limited information is available on the iron and manganese concentrations. This situation justifies the proposed study focused on collecting well water samples, analyze and gather additional information on the levels of manganese, arsenic, iron and other quality parameters of the groundwater used for drinking water.

Among other treatment system and processes, aeration followed by rapid sand filtration is commonly applied for the simultaneous or successive removal of iron and manganese. Sharma (2002) showed the mechanisms involved in adsorptive iron removal and means that further improve upon the iron removal processes in practice e.g. shortening ripening time, improving filtrate quality and reducing filter backwash frequency. A similar study on manganese removal is lacking, while there is a strong need from practice to improve this process. Frequently encountered problems in practice include: gradual loss of manganese removal efficiency and very long start-up periods when filters have been filled with new sand.

In addition, during extended field tests of the UNESCO-IHE – Family Filters for arsenic removal from groundwater, a problematic manganese removal has been observed. Initially the manganese removal started up slowly; contrary to expectation, after a couple of month's manganese release occurred. The manganese concentrations of the filtrate, exceeded that of the raw groundwater. This study therefore sets out to investigate the manganese adsorption capacities of various filter media, the effect of process conditions on the adsorption process and the oxidation kinetics of adsorbed iron (II) and manganese (II) and the factors that enhance these processes.

# 1.6    Research objectives

The specific objectives of this research are:

1. To screen the groundwater quality with highlights on the presence of manganese, arsenic, iron content in selected regions of the gold-belt zone of Ghana, and to identify the geological formations associated with the contaminated aquifers.
2. To determine manganese adsorption capacities of iron oxide coated sand under various process conditions.
3. To determine the effect of pH on adsorption capacities of selected media for manganese and to model the adsorption phenomenon.
4. To study the rate of adsorption of Mn (II) onto one or more selected media under different oxic conditions.
5. To study the effect of manganese and iron loading on the formation of a catalytic manganese oxide coating and the subsequent effect on the start-up of manganese removal in pilot rapid sand filters.
6. To study the release of manganese from filter media and investigate measures to optimize the performance of the UNESCO-IHE Family Filter in the removal of manganese, arsenic and iron when treating water with high ammonium content.
7. To investigate the oxidation kinetics of adsorbed iron and manganese at different aqueous pH values.

## 1.7    Outline of the thesis

This thesis has been organized into eight chapters. The current chapter gives a background to the dissertation including the problems associated with the use of groundwater as a resource for drinking water production, the chemistry, sources, mobilizations and removal technologies of contaminants like manganese, arsenic and iron. In-depth review of the various adsorptive mechanisms, the objectives to be addressed, the relevance and expected output of the study have been highlighted in this chapter.

Chapter 2 covers an inventory of the water quality of the existing wells in the gold-belt zone of Ghana focusing especially on the presence of manganese, arsenic, and iron. This chapter further identifies the communities covered by the study that are at risk of arsenic contamination through the use of geological information systems.

In chapter 3, the manganese adsorptive capacity of iron oxide coated sand and the effect of process conditions on the adsorption process are discussed.

Chapter 4 presents an investigation into the manganese adsorption of various selected filter media (both commercial and locally available media) and highlights the application of various kinetic models to the acquired bench scale experimental data.

Chapter 5 presents results of pilot plant studies conducted at the Haaren water treatment plant in the Netherlands to investigate the effect of manganese concentration and rate of filtration on the ripening and start-up of manganese removal in a rapid sand filter.

In chapter 6, the focus is on the release of manganese from filter media and probable conditions that facilitate the release. Measures to adopt in order to curtail the manganese release from filter were considered and used to optimize the performance of the Unesco-IHE Family Filter in the removal of arsenic, iron and manganese when treating water with high content of ammonium, manganese, iron and arsenic under laboratory conditions.

Chapter 7 covers the results of short column kinetic studies on the oxidation of adsorbed iron on iron oxide coated sand and adsorbed manganese on Aquamandix (manganese mineral) using modeled water under laboratory conditions.

Finally, Chapter 8 summarizes the results of this study and gives recommendations for practice and further research.

## 1.8    References

Abberantly, C. O., Ghanaian, E. V. 1993 Health effects of inorganic arsenic in drinking water production. *Journal, AWWA, WQTC, Miami, Fla.*, Nov. 7 – 11.

Abedin, M., Cotter-Howells, J. and Meharg, A.A. 2002. Arsenic uptake and accumulation in rice (Oryza sativa L.) irrigated with contaminated water. Plant and Soil, 240 (2), p:311 – 319.

Acharyya, S.K., Chakraborty, P., Lahiri, S., Raymahashay, B.C., Guha,S., Bhowmik, A., Chowdhury, T.R. Basu, G.K., Mandal, B.K., Biswas, B.K., Samanta, G. Chowdhury, U.K., Chanda, C.R., Lodh, D., Roy, S.L., Saha, K.C., Roy, S. Kabir, S., Quamruzzaman, Q., and Chakraborti, D. 1999 Enviroment: Arsenic poisoning in the Ganges delta. Nature, 401 (6753), p 545.

Alfe, D., Gillan, M.J., Price, G.D. 2003 Thermodynamics from first principles: temperature and composition of the Earth's core. Mineralogical Magazine 67 (1): 113 – 123.

AWWA (American Water Works Association) 1990 Water Quality and Treatment – A Handbook of Community Water Supplies. Tech. Editor – Pontius, F. W. p160-161, 699 – 701,.

Andreae, M. O., Andreae, T.W. 1989. Dissolved arsenic species in the Schelde estuary and watershed, Belgium. Estuar. Coast. Shelf Sci. 29, 421 – 433.

Appelo, C. A. J. and Postma, D.: 1994 Iron in groundwater. In 'Geochemistry, groundwater and pollution.' A. A. A. Balkena Publishers, Rotterdam, Netherlands. p279 – 284.

Azcue, J.M., Murdoch, A., Rosa, F., Hall, G.E.M., 1994 Effects of abandoned gold mine tailings on the arsenic concentrations in water and sediments of Jack of Clubs Lake, BC. Environ. Technol. 15, 669 – 678.

Bajpai, S. and Chaudhuri, M.: 1999 Removal of arsenic from groundwater by manganese dioxide-coated sand. Journal of environmental engineering. Vol.125, No.8. 782 – 784.

BGS & DPHE : 2001, Arsenic contamination of groundwater in Bangladesh. Kinniburgh DG & Smedley PL ed. Vol 2: Final report. British Geological Survey Report WC/00/19 Keyworth, UK, British Geological Survey. pp 215.

Borgono, J.M., Vincent, P., Venturino, H.; Infante, A., and United States of America, N.I.O.E.H.S.A.S. 1977 Arsenic in the drinking water of the city of Antofagasta; epidemiological and clinical study before and after the installation of a treatment plant. United States of America, National Institute of Environment Health Sciences [Arsenic Symposium].

Burke, J.J. and Moench, M.H. 2000. Groundwater and society: resources, tensions and opportunities. United Nations Publication ST/ESA/205.

Castro de Esparza, M.L. 2006 The presence of arsenic in drinking water in Latin America and its effect on public health. In Proceedings of the International Congress on 'Natural Arsenic in Groundwaters of Latin America'. Mexico City, 20 – 24 June 2006.

Chakravarty, S., Dureja, V., Bhattacharyya, G., Maity, S. and Bhattacharjee, S. 2003. Removal of arsenic from groundwater using low cost ferruginous manganese ore. Water Research 36 (2002): 625 – 632.

Clarke, R.: Lawrence, A.-R.: Foster, S.D.1996. Groundwater – a threatened resource. Nairobi, United Nations Environmental Programme Environment Library 15.

Crerar, D. A.: Cormick, R.K.; Barnes, H.L. 1980 Geology and Geochemistry of Manganese. Varentsov I.M., Grasselly Gy. (editors). Vol.1. Stuttgart: E. Schweizerbart'sche Verlagsbuchhandlung; pp 293 – 334.

Crerar, D.A., and Barnes H.L. 1974 Deposition of deep-sea manganese nodules. Geochim. Cosmochim. Acta 38: 270 – 300.

Criaud, A., Fouillac, C., 1989. The distribution of arsenic(III) and arsenic(V) in geothermal waters: Examples from the Massif Central of France, the Island of Dominica in the Leeward Islands of the Carribean, the Valles Caldera of New Mexico, USA and southwest Bulgaria. Chem. Geol. 76, 259 – 269.

Crittenden, B. and Thomas, J. W.: (1998) 'Adsorption Technology & Design'. Publisher: Butterworth Heinemann – (The Spartan Press) Biddles Ltd, Guildford and King's Lynn.

Das, D., Chatterjee, A.,Mandal, B.K., Samanta, G., Chakraborti, D., and Chanda, B. 1995 Arsenic in groundwater in six districts of West Bengal, India: The biggest arsenic calamity in the World – Part 2. Arsenic concentration in drinking water, hair, nails, urine, skin-scale and liver tissue (Biopsy) of the affected people. Analyst – Letchworth, 120 (3), pp 917 – 924.

Dasgupta, D.R.: 1965 Mineral Mag. 35, 131-139.

Davison, J. 1991 Bio-carb and wetlands – Passive affordable acid mine drainage treatment: In: Proceedings, 12th Ann. West Virginia Surface Mine Drainage Task Force Symposium, Morgantown, WV, p11.

Demayo, A. 1985 Elements in the Earths Crust, In: CRC Handbook of Chemistry and Physics, 66th Edition. Ed. Weast, Robert C., CRC Press Inc., Boca Raton, FI, p F145.

Dong Li, Zhang, J., Wang, H. and Wang, B. 2005 Operational performance of biological treatment plant for iron and manganese removal. Journal of Water Supply: Research and Technology – AQUA 54:1.

Driehaus, W., Jekel, M., Hildebrandt, U., 1998 Granular ferric hydroxide – a new adsorbent for the removal of arsenic from natural water. J. Water Supp. Res. Technol. – Aqua 47, 30 – 35.

Driehaus, W., Seith, R., Jekel, M., 1995 Oxidation of arsenate (III) with manganese oxides in water-treatment. Water Res. 29, 297 – 305.

EAWAG news 49e December (2000) 'Ground water research in practice.' Information Bulletin of the EAWAG. Swiss Fedral Institute for Environmental Science and Technology. A Research Institute of the ETH-Domain. CH-8600 Duebendorf. p18-20.

Edmunds, W.-M. and Smedley, P.-L.1996. 'Groundwater, Geochemistry and Health'. In: J.D. Appleton, R. Fuge and G.J.H.McCall (eds.), Environmental Geochemistry and Health with Special Reference to Developing Countries. London, The Geological Society Publishing House.

Ellis D, Bouchard C. Lantagne G. 2000 Removal of iron and manganese from groundwater by oxidation and microfiltration. *Desalination* 2000; 130: 255-64.

Emerson D. 1989 Ultrastructural organization, chemical composition of the sheath of Leptothrix discophora SP-6 PhD Dissertation, Cornell University, Ithaca, New York.

Ehrlich[a], H. L, 1996 How microbes influence mineral growth and dissolution. Chem. Geol. 132: 5 – 9.

Ehrlich[b], H. L, 1996 Geomicrobiology, Marcel Dekker, New York – USA.

Edwards, M. 1994 Chemistry of arsenic removal during coagulation and Fe-Mn oxidation.J.Am. Wat. Wks.Assoc., 86(9),64.

European Union Council Directive 1998; European Union Council Directive on quality intended for human consumption.

Fact Sheet July 2001 ' What you need to know about manganese in drinking water http://www. dph.state.ct.us/publications/BCH/EEOH/manganese. Division of Environmental Epidemiology and Occupational health. Hartford, CT 06134 - 0308.

Ferguson, J.F. and Gavis, J. 1972 A review of the arsenic cycle in natural waters. *Water Res.* 6, 1259 - 1274.

Foster, S.D. –D.; Chilton, P.-J.; Moench, M.; Cardy, W.-F.; Schiffler, M. 2000: Groundwater in Rural Development: Facing the Challenges of Supply and Resource Sustainability. New York, World Bank Technical paper 463.

Gardner-Outlaw, T. and Engelman, R. 1997. Sustaining water , Easing Scarcity: A Second Update. Washington DC, Population Action International.

Gallard, H.U., von Gunten, U. 2002 Chlorination of natural organic matter: kinetics of chlorination and of THM formation. *Water Research* 36: 65-74.

Golden, D.C.; Dixon, J.B. and Kanehiro, Y.: 1993 Austr. J.Soil Res. 31, 51-66.

Ghosh, M. M. and O'Connor, J. T.: Engelbrecht, R. S. (1996) Precipitation of iron in Aerated Groundwater: *Journal of Sanitary Engineering Division*, ASCE; vol. 90, No SA1, p 199-213, paper 4687.

Ghiorse, W.C. 1984 Biology of iron and manganese – depositing bacteria. Annu. Rev. Microbiol. 38: 515 - 550.

(GWCL) Ghana Water Company Limited – At a glance: 2000 The Public Relation Department GWCL – GHANA.

Hiemstra, T., van Riemsdijk, W.H., 1996 A surface structureal approach to ion adsorption: the charge distribution, CD. Model. J.Colloid Interface Sci. 179, 488 – 508.

Hochella, M. F.and White, A. F. (eds): (1990) Mineral water interface geochemistry, Rev, in Mineral., v. 23, pp. 603.

http://www.mdgmonitor.org,2007 (Last accessed April, 2008).

Huang, P. M. 1991 Kinetics of redox reactions on manganese oxides and its impact on environmental quality. In: D.L. Sparks, D.L. Suarez (eds) Rates of Soil Chemical Processes, pp 191 – 230 Soil Science Society of America Madison, Wisconsin.

INEP (United Natios Environment Programme) 1999. Global Environmental Outlook 2000. London, earthscan Publications.

Jenne, E. A. 1968 In Trace Inorganics in Water; Gould, R.F., Ed.; ACS Advances in Chemistry, Vol. 73, pp 337 – 387.

King, W. D.: 1998 Role of carbonate speciation on the oxidation rate of Fe(II) in aquatic system. Environ. Sci. Technol. 32(19): 2997-3003.

Kolker, A. 2000 Arsenic content of pyrite in sedimentary aquifers: The Marshall Sandstone of southeastern Michigan, USA. In: Abstracts, pre-congress worshop (Bwo 10) proceedings. Arsenic in groundwater of sedimentary aquifers. 31st International Geological Congress, Rio de Janeiro, Brazil pp 65 – 68.

Manceau, A and Charlet, L. 1992 *J.Colloid Interface Sci.* 148: 443 – 458.

Mann, S. Sparks, N.H.C., Scott, G.H.E and deVrind-de Jong, E.W.: 1988 Oxidation of manganese and formation of $Mn_3O_4$ (Hausmannite) by spore coats of a Marine Bacillus sp. Applied and Envronmental Microbiology. Vol. 54. No.8, 2140 - 2143.

Manning, B.A., Goldberg, S. 1996. Modeling competitive adsorption of arsenate with phosphate and molybdate on oxide minerals. Soil Sci. Soc. Am. J. 60, 121 – 131.mdg monitoring: http://www.mdgmonitor.org,2007 (Last accessed – April 2008).

Mattson, J. S. and Mark, Jr. H. B.: 1971 Activated carbon: surface chemistry and adsorption form solution, Marcel Dekker, Inc. N.Y.

Mouchet, P. 1992, From conventional to biological removal of iron and manganese in France. *J.Am Water Works Assoc.* 84(4): 158-66.

Millero, F. J. 1990 *Mar. Chem.* 30. 205 – 229.

Millero, F. J. and Hawke, D. J.: 1992 *Mar. Chem.* 40. 19 – 48.

Moore, J.N., Walker, J. R. and Hayes, T.H. 1990, Reaction scheme for the oxidation of As (III) to As(V) by birnessite. Clays and Clay Minerals 38(5): 549 – 555.

Mostert, E. 2006 Ingrated water resources management in the Netherlands: How concepts Function. *Journal of Contemporary Water Research & Education* (135) 19 - 127. UCOWR.

Murray, J.W. 1974 The surface chemistry of hydrous manganese dioxide. J.Coll. Interface Sci. 46, 357 – 371.

Nealson, K.H.1983 The mikcrobial manganese cycle. In W.E.Krumbein (ed). Microbial geochemistry. Blackwell Sci. Publishers, Oxford.

Nealson, K. H.; Tebo, B.M. and Rosson, R. A. 1988 Occurrence and mechanisms of microbial oxidation of manganese. Adv. Appl. Microbiol. 33: 279 – 318.

Nealson, K.H. Rosson R. A. and Myers, C.R. 1989 Mechanism of oxidation and reduction of manganese. In:TJ Beveridge, RJ Doyle (eds) Metal Ions and Bacteria, pp 383 – 411 John Wiley & Sons, New York.

Nickson R., McAthur J., Burgess W., Ahmed K.M., ravenscroft P. and Rahman M.: 1998, Arsenic poisoning of Bangladesh groundwater (letter). *Nature*, 395(6700): 338.

Nimick, D.A., Moore, J.N., Dalby, C. E., Savka, M.W., 1998 The fate of geothermal arsenic in the Madison and Missouri Rivers, Montana and Wyoming. Water Resour. Res. 344, 3051 – 3067.

Nordstrom, D.K.: 2002 Worldwide Occurrences of Arsenic in groundwater. – Public Health. Science 21. Vol. 296. No. 5576, pp. 21243 – 2145.

O'Connor, J.T. 1971 Iron and Manganese: Water quality and treatment – A handbook of Public Water Supplies. Chap. 11, pp 378 – 396. McGraw-Hill, New York.

Oscarson, D.W., Huang, P.M., Liaw, W. K. and Hammer, U.T. 1983 Kinetics of Oxidation of Arsenite by various manganese dioxides. Soil Science Society of America Journal 47(4):644 – 648.

Ostwald, J. 1984 Mineral. Mag. 48. 383-388.

Petrusevski, B.; Sharma, S.K.; Krius, F.; Omeruglu, P. and Schippers, J.C. 2002 Family filter with iron-coated sand: solution for arsenic removal in rural areas. *J. of Wat. Sc. and Technol. Water Supply*. 2,(5 – 6) 127 – 133.

Post, J.E. 1999 Manganese oxide minerals: crystal structures and economic and environmental significance. In: Proc. Natl. Acad. Sci. USA. Vol. 96. pp 3447 – 3454. Colloquium paper.

Ramstedt M. 2004 Chemical processes at the water-mangnite ($\gamma$-MnOOH) interface. PhD thesis pp 6, 13. Dept. of Chemistry; Inorganic Chemistry, Umeå University.

Reijnders, H.F.R.; van Drecht, G.; Prins,H.F. and Boumans, L.J.M. 2007 The quality of the groundwater in the Netherlands. *Journal of Hydrology* 207(1998), 179 – 188.

Scott, M.J. and Morgan, J.J. 1996 Reactions at oxides surfaces. – Oxidation of Se(IV) by synthetic birnessite. Environ. Sci. Technol. 30, 1990 – 1996.

Schott, J., Berner, R. A.:1983 X-ray photoelectron studies of the mechanism of iron silicate dissolution during weathering. Geochim. Cosmochim. Acta v. 47, pp. 2233-2240.

Sharma, S.K., Greetham, M.R. and Schippers J.C. 1999 Adsorptive iron (II) onto filter media. *Journal of Water Supply: Research and Technology – AQUA* 48 (3), 84 - 91.

Shiller, A. M. 2004 Temperature dependence of microbial Mn oxidation as a control on seasonal concentration variability of dissolved metals in flood-plain rivers. In Proc. Of the 227th National American Chemical Society Spring Meeting, Anaheim, CA. 28 March – 1st April. 2004.

Sikora, F.J.; Behrends, L.L.; Brodie, G. A. and Taylor 2000. Design criteria and required chemistry for removing manganese in acid mine drainage using subsurface flow wetlands. Water Environ. Res. 72(5): 536 – 544.

Slejko, F. L. (editor) 1985 Adsorption Technology: a step-by- step approach to process evaluation and application. Chemical Industries / 19. Tall Oak publishing, Inc. Voorhees, New Jersey. Pp 1 -30.

Smedley, P. L. and Kinniburg, D. G., 2002 A review of the source , behaviour and distribution of arsenic in natural waters. Applied Geochemistry 17 (2002) 517 – 518, 524.

Sracek, A., Bhattacharya, P.,Jacks, G. and Gustafsson, P.J.: 2001 Mobility of arsenic and geochemical modeling in groundwater environment. In: Jacks, G., Bhattacharya, P and Khan, A.A. (eds.) Groundwater Arsenic Contamination in the Bengal Delta Plain of Bangladesh. Proceedings of the KTH-Dhaka University Seminar, KTH Special Publication, TRITA-AMI Report 3084, pp 9 – 20.

Stenkamp, V.S. and Benjamin, M. M.: 1994 Effect of iron oxide coatings on sand filtration: *Journal of AWWA*; vol. 86, No. 8, p 37-50.

Stone A.T. Ulrich, H. J. 1989 Kinetics and reaction stoichiometry in the reductive dissolution of manganese(IV) dioxide and Co(III) oxide by hydroquinone. J. Colloid Interface Sci 132:509 – 522.

Stumm, W. and Lee, G.G. 1961 Oxygenation of ferrous iron: Industrial Engineering and Chemistry. Vol. 53, No. 2, pp 143 – 146.

Stumm, W. and Morgan, J.J. 1996 Aquatic Chemistry, chemical equilibria and rates in 3rd; Wiley; New York.

Sunda W.G. Kieber, D. J. 1994 Oxidation of humic substances by manganese oxides yields low-molecular-weight organic substrates. Nature 367: 62 – 64.

Sung, W. and Morgan, J.J. 1980 Kinetics and products of ferrous ion oxygenation in Aquatic systems. Environmental Science and Technology. Vol. 14, No.5, pp 561 – 568.

Takai, T. (1973) Studies on the Mechanism of catalytic defferrization of Fe(II): *Journal of Japan Water Works Association*, No 446, p22-23.

Tamura, H. Goto, K. and Nagayama, M. 1976 The effect of ferric hydroxide on the oxygenation of ferrous ions in neutral solutions: *Corrosion Science*; vol. 16, p197-207.

Tebo, B.M. 1991 Manganese (II) oxidation in the suboxic zone of the Black Sea. Deep –Sea Res. 38: S883 – S905.

Tebo, B.M. 1995 Metal precipitation by marine bacteria: potential for biotechnological applications. In: JK Setlow (ed) Genetic Engineering – Principles and Methods 17:231 – 263, Plenum Press, New York.

Tebo, B.M., Ghiorse, W. C., van Waasbergen, L.G., Siering, P.L., & Caspi, R. 1997 In Geomicrobiology: Interactions between Microbes and Minerals, eds Banfield, J. F. & Nealson, K. H. (Mineral. Soc. Am., Washington, DC), ACS Advances in Chemistry Series, Vol. 73, pp, 337-387.

Theis, T.L. and Singer, P.C. 1974 Complexation of iron (II) by organic matter and its effect on iron (II) oxygenation. *Env. Sci. Technol.* 8 (6), 569 – 573 .

Turner, S. and Buseck, P. R. 1979 Manganese oxide tunnel structures and their intergrowths. Science vol. 203, No. 4379, p 456 – 458.

Tufekci, N. and Sarikaya, H.Z. 1996 Catalytic effects of high Fe (III) concentrations on Fe (II) oxidation. *Wat. Sci. Tech.* Vol. 34 No. 7 – 8. 389 – 396.

USEPA, 1993, 'Office of groundwater & drinking water. Arsenic in drinking water treatment technologies', http://www.epa.gov/safewater/ars/treat.html. (Last access – April 2008).

USEPA 1994 Drinking water criteria document for manganese. Washington, DC, US Environmental Protection Agency, Office of Water.

Van Vliet, B. M. and Weber. Jr.: 1981 Comparative performance of synthetic adsorbents and activated carbon for specific compound removal from wastewaters. *J. Water Poll. Control Fed.*, 53, 11, 1585 – 1598.

Water in the Netherlands; 2005 & 2006 http://www.waterland.net/index.cfm.site (Last accessed August 2008).

Water Quality Assessments 1997 A guide to the use of biota, sediments and water in environmental monitoring. 2$^{nd}$ Edition. Edited by Deborah Chapman – UNESCO, WHO, UNEP. Publishers:- Chapman and Hall.

Water resources: FAO: AQUASTAT 2002 land and population : FAOSTAT, except for the United States(Conterminous, Alaska and Hawaii) US Census Bureau.

Wehrli, B. 1990 In Aquatic Chemical Kinetics; Stumm, W., Ed.; Wiley: New York, pp311 – 337.

Wehrli. B, Friedl, G. and Manceau, A. 1995 Reaction rates and products of manganese oxidation at the sediment-water interface. In: C.P. Huang, C. R. O'Melia, J.J. Morgan (eds) Aquatic Chemistry: Interfacial and Interspecies Processes. American Chemical Society, Washington, DC.

Welch, A. H., Lico, M.S. Hughes, J.L. 1988 Arsenic in ground-water of the Western United States. Ground Water 26, 333-347.

Weber, W. J. Jnr. 1972 Physicochemical processes for water quality control: Wiley Interscience Publication: John Wiley & Sons Inc. U.S.A.

Weber, W. J., Jr., and J.C. Morris: 1963 Kinetics of adsorption on carbon from solution,' *Jour. San. Eng. Div., Am. Soc. Civil Eng.*, 89, SA 2: 31- 59 'Closure' *Jour. San. Eng. Div., Am. Soc. Civil Eng.*, 89, SA6: 53-55.

White, A. F. 1990 Heterogenous electrochemical reactions associated with oxidation of ferrous oxide and silicate surface in Hochella, M. F. and White, A. F. (eds), 1990, Mineral water interface geochemistry, Rev, in Mineral., v. 23, p. 447 – 509.

Wilkie, J.A., Hering, J.G. 1998 Rapid oxidation of geothermal arsenic(III) in streamwaters of the eastern Sierra Nevada. Environ. Sci. Technol. 32, 657 – 662.

Wilson, S.D.; Kelly, W.R.; Holm, T.R., and Talbott, J.L. 2003 Arsenic removal in water treatment facilities: survey of geochemical factors and pilot plant experiments. Midwest

Technology Assistance Centre for small public water systems (MTAC final report). Illinois - USA. http://mtac.sws.uiuc.edu/mtacdocs/MTAC2003AnnualReport.pdf (Last accessed – June 2008).

World population prospects: The 2006 Revision United Nations – Department of Economic and Social Affairs: http://www.en.wikipedia.org/wiki/Africa (Last accessed – April 2008). World Water Assessment Programme: Water for People, Water for life – The United Nations.

World Water Development Report: UNESCO-WWAP 2003. A joint report by the twenty- three UN agencies concerned with freshwater. WHO, 1996: Guidelines for Drinking water quality vol. 2 , WHO, Geneva.

WHO, 1999: Arsenic contamination in the world.
   http://www.Who.int/peh-super/Othlec/Arsenic/Series2/003.html
   (Last accessed – June 2004).

WHO, 2006 Guidelines for drinking – water quality. First Addendum to Third edition. Vol. 1 recommendations 186 & 398, Geneva.
   http://www.who.int/water_sanitation_health/dwq/gdwq0506.pdf
   (Last accessed – April 2008).

WHO, Safer Water, Better Health report. 2008 Costs, benefits and sustainability of interventions to protect and promote health. A. Pruss-Ustun, R. Bos, F.Gore, J. Bartram.
   http://whqlibdoc.who.int/publications/2008/9789241596435_eng.pdf
   (Last accessed - September, 2008).

WWF (World Wide Fund for Nature). 1998. Living Planet Report 1998: Over consumption is Driving the Rapid Decline of the World's Natural environments. Gland, Switzerland. Vewin 2007: http://www.vewin.nl; 2008 (Last accessed – May 2008).

Vrba, J. and Zaporozec, A. 1994. Guidebook on Mapping Groundwater Vulnerability. International Contributions to Hydrogeology. Wallingford, International Association of Hydrological Sciences.

Yan-Chu (1994) 'Arsenic distribution in soils.' Arsenic in the environment. Part 1: Cycling and characterization, J.O.Nriagu ed., Wiley, New York, 17 – 49.

Zektser, I. and Margat, J. 2003 Water for people, water for life; Part II – A look at the World's freshwater resources. 1$^{st}$ Report. UNWWD Report. pp 78. Paris, UNESCO: http://www.unesco.org/water/wwap/wwdr/wwdr1/pdf/chap4.pdf (Last accessed - June 2008).

Zhang, J. Lion, L.W. Lion, Nelson, Y.M.: Shuller and Ghiorse, W.C. 2002, Kinetics and Mn (II) oxidation by Leptothrix discophora SSI. Goechim. Cosmochim. Acta 65(5): 773-7781.

# CHAPTER TWO

# PRESENCE OF MANGANESE, ARSENIC AND IRON IN GROUNDWATER WITHIN THE GOLD-BELT ZONE OF GHANA

Main part of this chapter was published as:
Buamah, R. Petrusevski, B. and Schippers, J.C. (2008) Presence of arsenic, iron and manganese in groundwater within the gold-belt zone of Ghana. *Journal of Water Supply*: *Research and Technology* – AQUA 57.7: 519 – 529.

# Abstract

About 45% of the total drinking water in Ghana is produced from groundwater. The presence of manganese and arsenic in groundwater above the recommended WHO drinking- water guidelines pose a threat to consumers' health. In addition, manganese together with iron poses an aesthetic challenge in water when present in high concentration (i.e. 0.1mg/l for manganese and 0.3 mg/l for iron). They confer bad taste to the water and stain laundry and plumbing fixtures. To provide additional information on groundwater quality in the gold-belt zone of Ghana, nearly 290 well water samples from three regions namely Ashanti, Western and Brong-Ahafo, were collected and analyzed for presence of manganese, iron and arsenic. Thirteen percent of the wells in Ashanti and 29% in the Western region exceeded 0.4 mg/l – the WHO health-based guideline value for manganese. Brong-Ahafo, Ashanti and Western, regions had 5%, 35%, and 50%, of wells, respectively with iron levels above 0.3 mg/l, the Ghana drinking-water guideline value commonly accepted for iron.

It was found that 5 – 12 % of all the sampled wells had arsenic levels exceeding the 10 µg/l - WHO provisional guideline value. It is estimated that between 500 000 – 800 000 inhabitants in the communities covered in this study use untreated water with [As] > 10 µg/l. Communities within the studied area with high arsenic presence in their groundwater are located within the Birimian and Tarkwaian geological formations. Most of these arsenic contaminated wells (70%) have been in use for more than 15 years. Elevated arsenic levels in the aquifers are presumably due to arsenopyrite oxidation and reductive dissolution and desorption of ferric oxides.

**Keywords**: arsenic, Ghana, gold-belt, groundwater, iron, manganese,

# 2.1    Introduction

Groundwater is in general of good and constant quality and therefore its usage for drinking water production is associated with relatively low capital, operational and maintenance costs, and is consequently a very attractive drinking water resource for scattered rural communities in developing countries. In recent times there has been growing concern about the presence of heavy metals like arsenic, iron and manganese in drinking water sources because of their health and / or aesthetic implications. Inorganic arsenic species are carcinogenic to humans. A provisional guideline of 10 µg/l for arsenic has been proposed (WHO, 1996, 2004). No health related guideline has been proposed for iron in drinking water. For manganese, a health-based guideline of 0.4 mg/l has been proposed (WHO, 2004).

Iron and manganese impact negatively the aesthetic value of drinking water. When groundwater contains high iron concentrations, people in rural communities tend to rivet to contaminated surface water sources for household activities. The WHO (1996) and the Ghana Standard Board's guidelines for drinking water quality for iron, manganese and arsenic in piped water supply systems are quoted as < 0.3 mg/l for iron, 0.1 mg/l for manganese (for aesthetic reasons) and 10μg/l for arsenic (GSB, 1998).

## 2.1.1 Sources of manganese, iron and arsenic in groundwater

Manganese normally does not occur as a free metal in nature but mostly in the form of oxides, sulfides, carbonates and silicates (Post, 1999). One major cause of manganese mobilization in aquifers is reductive decomposition and dissolution of compounds such as Mn-OOH and $MnO_2$. In the normal pH range of groundwater (pH 5 – 8), dissolved iron is present as $Fe^{2+}$. The main sources of $Fe^{2+}$ include:

- the dissolution of iron (II) bearing minerals;
- the reduction of iron oxyhydroxides (Fe-OOH) present in the sediments e.g. magnetite, ilmenite, pyrite, siderite, iron (II) bearing silicates and clay minerals such as smectites (Appelo and Postma, 1994);
- the oxidation of arsenopyrites.

Arsenic is present throughout the earth's crust in various compounds, the most common being arsenopyrite (FeAsS). It is widely distributed and transported in the environment by water (Chien-Jen Chen et al. 1999). Anthropogenic activities like geothermal drilling, coal mining and smelting enhance the proliferation of arsenic compounds in the environment. High arsenic concentration in groundwater most commonly results from chemical processes including dissolution and desorption from oxyhydroxides and arsenic – bearing minerals, arsenopyrite oxidation, upflow of geothermal water and evaporative concentration etc. as elaborated in chapter 1.

## 2.1.2 Simplified geology of Ghana

Most parts of Ghana lies within the Precambrian Guinea Shield of West Africa. Ghana can be subdivided geologically into three different major parts:

- south-west and north-western parts which are composed of the oldest crystalline rocks of the Birimian and Tarkwaian formations with common major intrusions of granite;
- south-eastern part, incorporating the Akwapim-Togo Hill Range consists of Togo Series with quartzite and shale; the Buem Group which consists of a mixed sequence of metamorphosed sedimentary and igneous rocks and the Dahomeyan System comprising gneisses;
- central and eastern part (Voltaian and Highlands Basin) are composed of consolidated sandstones, mudstones and limestones (Kesse, 1985).

About two-thirds of Ghana's geology is dominated by the Birimian system (Figure 2.1). The Birimian is the most mineralized formation and comprises metamorphic sediments with five parallel evenly spaced, volcanic belts (Leube et al., 1990). Between the volcanic belts of the Birimian system are intervening basins that are filled by sediments. The Birimian intrusive granitoids of this formation contain secondary discontinuities like fractures, joints etc., that are of importance in the formation of permeable groundwater reservoirs. Wells of average depth 60 m drilled in the Birimian formation typically yield about $2 - 36$ m$^3$/hour (Mining portal of Ghana, 2006.).

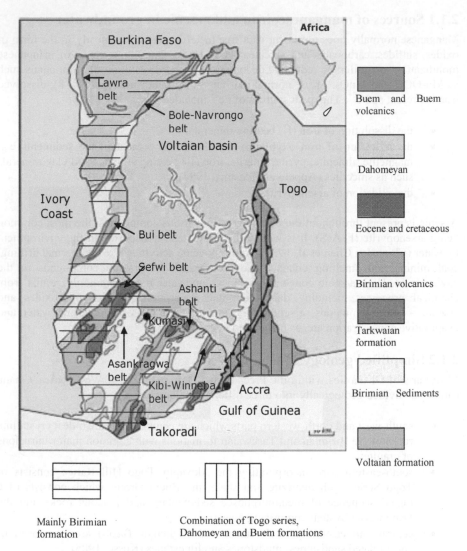

**Figure 2.1** Simplified geological map of Ghana showing the gold-belts.
(Adopted and modified from the Mining portal of Ghana:- http:www.ghana-mining.org.)

The Tarkwaian formations, derived and developed in the central portions of the Birimian gold belts are made up of meta-sediments i.e. conglomerates, sandstones, quartzite and shale. They rest unconformably on the Birimian and contain openings along joint, bedding and cleavage planes and are good suppliers of groundwater (Leube et al 1990). The average yield from wells in this formation is about 9 m$^3$/ hour.

The Togo Series outcrops in the eastern part of the country and form a range of mountains and hills within the northeastern portion of Ghana (Kesse, 1985). The Buem formation that outcrops to the east of the Voltaian is made up of calcareous, argillaceous, sandy and ferruginous shales, sandstones arkoses, greywacke, and conglomerates etc.

The Dahomeyan System is considered to be the oldest formation in the country. The Dahomeyan System outcrops to the southeastern and northeastern portions of the country and comprises acidic and basic gneisses (Kesse, 1985). The impervious and massive nature of the Dahomeyan limits the available groundwater. Successful wells in this formation yield $0.5 - 11$ m$^3$/hour.

The Voltaian formation comprises the consolidated sedimentary rocks. The rock types are mainly horizontal sediments of sandstones, shales, arkoses, mudstones, sandy, limestone etc. The wells drilled in the Voltaian have average yield of $3.6 - 9$ m$^3$/hour. Mainly sedimentary rocks of the coastal block fault stretches from the southwest end to the southeast corner of the country.

## 2.1.3 The gold belts of Ghana

The gold-belt zone of Ghana extends for about 250 km across the country with a northeast to southwest strike (Figure 2.1). The mineralization extends to a depth of about 1250 m and has a lateral thickness in the order of 50 m (Smedley et al., 1996). The Birimian and Tarkwain formations are the main host rocks for deposits of gold and other precious minerals eg. diamond, bauxite etc. The vast majority of the Birimian gold deposits occur aligned along the flanks of the volcanic belts. The belts normally have a width of $15 - 40$ km and are separated from one another by a distance of about 90 km (Leube et al. 1990). The gold belts include Ashanti belt, Asankragwa belt, Sefwi belt etc. (Figure 2.1). Almost all of the major gold mines, abandoned or working, occur in the Kumasi basin in the southwest Ghana. These mines are concentrated mainly along the flanks of the Ashanti or Sefwi belt.

### 2.1.3.1 Are aquifers in the goldbelt-zone at risk of arsenic contamination?

The gold deposits in Ghana normally occur in close association with sulphide minerals especially arsenopyrite. Groundwaters in the gold-belt zone of Ghana are potentially vulnerable to the presence of elevated concentrations of dissolved arsenic as a result of the oxidation of the sulphide minerals. A perusal of the general groundwater quality data of Ghana has given indications of high levels of iron and manganese in some regions of the country but limited information on the levels of arsenic is available. Smedley et al. (1996) and Norman and Miller (1999) had reported arsenic levels of up to 141µg/l in groundwater in Bolgatanga in Northern Ghana and up to 64 µg/l in the largest gold mining town of Obuasi. Other communities within the gold belt zone of the country still remain to be investigated. Presently in Ghana there has been an upsurge of incidence of cancer and no particular cause has been established that is associated with this current trend (Nkyerkyer, 2000). Consumption of drinking water with elevated arsenic level may be one of the causes.

### 2.1.3.2 Use of groundwwater for water supply in Ghana

At the moment the water demand in the country stands at about 60 – 75 litres/p/day for the inhabitants in the big cities and 20 – 30 litres/p/day for inhabitants in small towns or communities. The big cities in the country with population size > 500 000 are normally supplied with treated water from surface water treatment plants. In this study however the focus was on the drinking water sources of the small communities. Most of the small towns depend on wells for their drinking water supply. On the average, one well is provided per a population of 300 people. Presently, 45% of the total drinking water production in Ghana is from groundwater and majority of the wells is used directly untreated as drinking water. The total number of wells in the Ashanti, Brong-Ahafo and Western regions that are the focus of this study is estimated to be between 6000 – 7000.

The objective of this study was to provide additional information on the arsenic, iron and manganese content of groundwater of Ashanti, Western and Brong-Ahafo regions in the gold-belt zone of Ghana. Other water quality parameters such as pH, conductivity, hardness of the groundwater in the regions were also determined.

## 2.2    Materials and methods

### 2.2.1 Communities within regions and districts covered by sampling

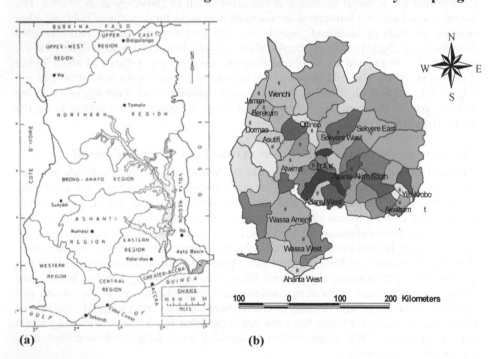

(a)                                                        (b)

**Figure 2.2** Map of Ghana showing (a) the regions and (b) the districts covered by the study (Figure 2.2A adopted from the Mining portal of Ghana, 2006).

The initial survey of the districts included in this study demonstrated no accurate inventory of the wells. Several non-governmental agencies and individuals have drilled wells within the communities without proper documentation. Communities included in the study were chosen to cover the various geological formations within the studied regions (see Figures 2.1 & 2.2). On the whole water samples from 286 wells (of depth 40 – 60 m) in 238 communities in Ashanti, Western and Brong-Ahafo regions were taken. Wells from 8 communities in Eastern region that do not fall within the gold-belt zone of the country were also sampled to provide insights into the groundwater quality outside of the gold-belt zone (Table 2.1).

**Table 2.1** Regions and districts included in the study.

| REGIONS | DISTRICTS | Number of communities sampled | Number of well samples |
|---|---|---|---|
| Ashanti (114) | Adansi West *** | 28 | 32 |
| | AAS** | 7 | 10 |
| | Atwima | 11 | 14 |
| | Offinso | 18 | 22 |
| | BAK* | 24 | 31 |
| | Sekyere East & West | 5 | 5 |
| Western (90) | Wassa West | 23 | 32 |
| | Wassa Amenfi | 27 | 34 |
| | Ahanta West | 20 | 24 |
| Brong-Ahafo (82) | Berekum | 15 | 16 |
| | Jaman | 8 | 8 |
| | Asutifi | 18 | 21 |
| | Wenchi, | 18 | 19 |
| | Dormaa | 16 | 18 |
| Eastern (8) | Akwapim North | 5 | 5 |
| | Yilo Krobo | 3 | 3 |
| TOTAL | | 246 | 294 |

Numbers in parentheses represent the total number of samples taken in that region
BAK* – Bosumtwe-Atwima-Kwawoma      Adansi West*** - The Obuasi area
AAS** - Asante Akim South

## 2.2.2 Analysis of samples

Acid-washed one-litre plastic containers were used for the sampling. For each well, two samples were taken: one had 3 ml of 1N $HNO_3$ acid added whereas the other sample was kept without any acid addition. The pH of the raw water samples were determined on site with WTW model pH340 - pH meter Using the acidified water samples the iron and manganese concentrations were determined with an atomic absorption spectrometer - Perkin Elmer AA100 with detection limits of 0.01 and 0.02 mg/l, respectively. For arsenic, the atomic absorption spectrometer (Perkin Elmer Analyst 5100) equipped with graphite furnace atomizer and HGA 300 programmer with arsenic detection limit of 4µg/l was used. The metallic ions analysis was carried out at the UNESCO-IHE laboratory in the Netherlands according to Dutch Standard Method NEN 6457.

## 2.3    Results and discussion

Most wells having high iron content often have some amount of manganese. These two metals (Mn and Fe) have some similar chemical reactions and both have their oxidised ions affecting the quality of water aesthetically, therefore the analytical results of their occurrences in the sampled wells have been discussed together.

### 2.3.1 Categorization of wells

The presence of the contaminants (Mn, Fe and As) was analyzed per district of the regions included in the study. Based on the measured contaminant concentrations in the groundwater, the wells were grouped in three categories for each contaminant (Table 2.2).

**Table 2.2** Classification of wells based on measured contaminant concentrations.

|  | CATEGORIES | | |
|---|---|---|---|
| *CONTAMINANTS* | Category 1 | Category 2 | Category 3 |
| Mn | Mn ≤ 0.1mg/l | 0.1 mg/l < Mn ≤ 0.4mg/l | Mn ≥ 0.4 mg/l |
| Fe | Fe ≤ 0.3 mg/l | 0.3 mg/l < Fe ≤ 1.0 mg/l | Fe ≥ 1.0 mg/l ** |
| As | As ≤ 10 µg/l | 10 µg /l < As ≤ 50 µg / | As ≥ 50 µg /l * |

   \* 50 µg/l is the current guideline value for arsenic in drinking water for many
      countries e.g. Bangladesh

  \*\* 1.0 mg/l is the iron concentration at which consumers could start complaining

### 2.3.2 Occurrence of manganese and iron

#### 2.3.2.1 Ashanti region

Thirteen percent of sampled wells in the region had manganese levels exceeding 0.4 mg/l, the WHO, health based guideline. Approximately 35% of the wells in the region tested had iron content >0.3 mg/l while 18% exceeded 1.0 mg/l. On the whole wells tested in the Asante Akim South district (AAS), followed by the Atwima had higher iron content than the other districts (Figure 2.3a).

#### 2.3.2.2 Western region

The Western region was found to have the highest percentage of wells (29%) with manganese levels exceeding 0.4 mg/l, the WHO health based guideline (WHO, 2004), (Figure 2.3b). The situation of Atwima and Wassa West districts is more precarious in the sense that 21% and 50% of their total number of wells respectively exceeded the

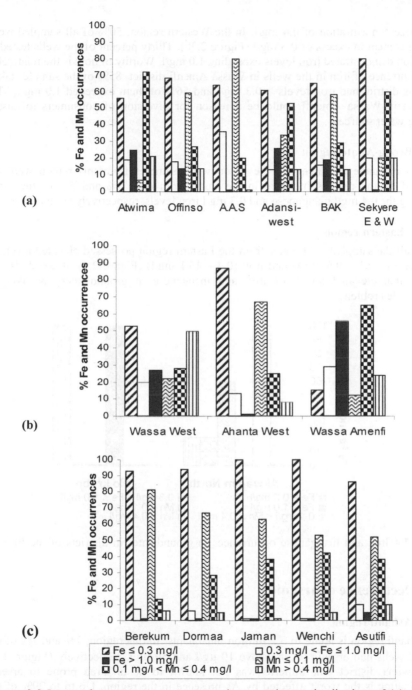

**Figure 2.3** Iron and manganese occurrences in groundwater in districts of (**a**) Ashanti (**b**) Western (**c**) Brong-Ahafo.

manganese concentration of 0.4 mg/l. In the Western region, 50% of all sampled wells had iron content in excess of 0.3 mg/l (Figure 2.3b). Thirty percent of the wells tested in this region demonstrated iron levels exceeding 1.0 mg/l. Worthy of note is the relatively high occurrence of iron in the wells in Wassa Amenfi district; 85% of the samples taken from this district had iron levels > 0.3 mg/l and 56% of them exceeded 1.0 mg/l. This condition in Wassa Amenfi could be a reason for diversion of consumers to unsafe drinking water sources.

### 2.3.2.3 Brong-Ahafo region

Iron and manganese contents in the groundwater samples from this region were the lowest of all the regions included in the study, with only 6% and 5% of the wells exceeding the 0.4 mg/l manganese and 0.3 mg/l iron levels, respectively (Figure 2.3c).

### 2.3.2.4   Eastern region

Almost all the sampled well water from the Eastern region possessed elevated levels of manganese ($\leq 0.1 - 0.9$ mg/l) and iron ($0.4 - 15.1$ mg/l) (Figure 2.4). It could also be inferred that elevated levels of iron and manganese in groundwater are likely a countrywide problem.

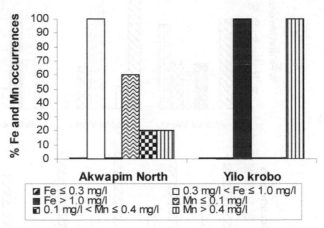

**Figure 2.4** Iron and manganese occurrences in groundwater in districts of the Eastern region.

## 2.3.3 Occurrence of arsenic

### 2.3.3.1 Ashanti region

The Ashanti region is home to 3.2 million inhabitants. In this region, 7% and 1% of the sampled wells had arsenic levels above 10 µg/l and 50 µg/l, respectively (Figure 2.5). The Atwima district that hitherto was not known to be an area prone to arsenic contamination is the worst affected by As presence in the region. Up to to 29% of the sampled wells in the Atwima district were found to have arsenic levels higher than 10 µg/l (Figure 2.5a). Wells in these communities have been in use as drinking water source

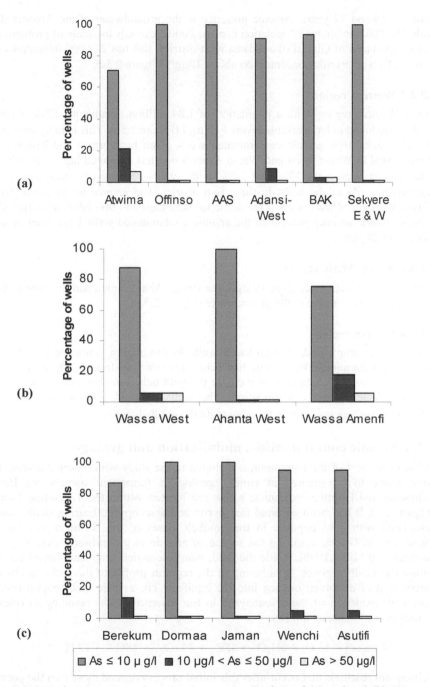

**Figure 2.5** Arsenic occurrences in ground water in districts of (a) Ashanti (b) Western and (c) Brong-Ahafo regions.

between 19 and 34 years. Arsenic presence in the groundwater of the Atwima district with 215 000 inhabitants (Population census, 2000), exceeds the scale of problem in the well-known area of Obuasi (the Adansi west district) that has 222 000 inhabitants and 9 % of wells with arsenic concentration above 10µg/l (Figure 2.5a).

### 2.3.3.2 Western region

In the Western region, with a population of 1.84 million inhabitants, 12% of sampled wells were found to have arsenic levels >10 µg/l (Figure 2.5b). This is in agreement with an estimate of 10% arsenic contaminated wells given by Norman and Miller (1999). Wells tested in Wassa West and Wassa Amenfi districts exhibited higher arsenic levels than Ahanta West with 13% and 24% of the wells respectively exceeding 10 µg/l. Remarkably, 6% of the wells in these two districts had arsenic levels above 50 µg/l. Wassa West and Wassa Amenfi are home to more than 450 000 inhabitants (population census, 2000). Seventy percent of the arsenic contaminated wells have been in use for more than 20 years.

### 2.3.3.3 Brong–Ahafo region

With 5% of its wells exceeding 10 µg/l, the Brong-Ahafo region had the least incidence of arsenic contamination in the groundwater (Figure 2.5c).

### 2.3.3.4 Eastern region

None of the samples taken from the 8 wells in this region contained any detectable arsenic concentration. These wells had acidic pH (varying from 5.2 to 6.9) similar to those of the arsenic contaminated wells in the gold belt zone, however, none contained any detectable arsenic contamination. The wells sampled here occur in geological formations without gold deposits and outside the goldbelt zone.

## 2.3.4 Arsenic contamination, mobilization and geology

Using GIS, most of the communities included in the study with elevated arsenic levels were found to be situated in similar geological formations namely the Birimian sediments and Birimian volcanics; a few are located within the Tarkwaian formation (Figure 2.6). It had been reported that pyrite and arsenopyrite bearing shale, normally associated with gold deposits in the goldbelt zones of the Birimian and Takwaian formations in Ghana, could be the source of arsenic in groundwater (Smedley, 1996; Norman and Miller,1999). Aside the gold, manganese-rich sediments have been found within the goldbelt zones. Weathering of the regolith profile of the rocks results in the percolation of dissolved oxygen into the aquifers. The arsenic mobilization processes start with oxidation of the arsenopyrite in the basement rocks resulting in release of ferrous and arsenate, as follows:

$$4 \text{ FeAsS} + 13 \text{ O}_2 + 6\text{H}_2\text{O} \rightarrow 4\text{Fe}^{2+} + 4\text{AsO}_4^{3-} + 4\text{SO}_4^{2-} + 12 \text{ H}^+ \qquad (2.1)$$

Subsequent reactions that occur after this initial process depend mostly on the prevailing redox conditions and the pH of the groundwater. Under oxidizing conditions the ferrous ions produced may be oxidized to ferric oxides or hydroxides that may (partially) adsorb the arsenic oxy- anion and thereby restrict arsenic mobility in the groundwater.

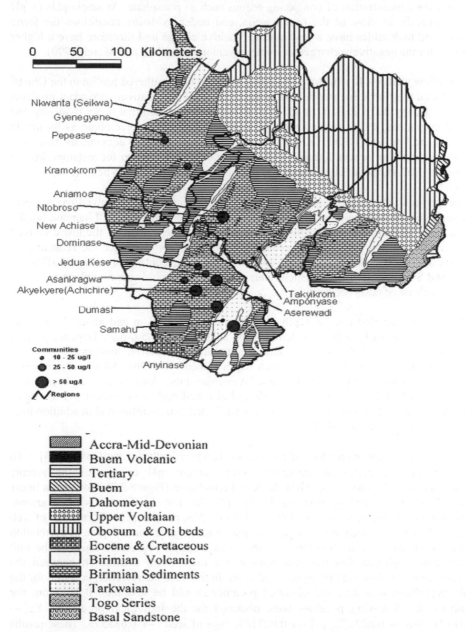

**Figure 2.6** Geological map of sampled regions showing communities with high arsenic contamination (ranges of arsenic concentrations in communities included in the study are depicted with circular spots).

The extent of adsorption is dependent on the arsenic speciation, arsenic concentration, pH and the concentration of competing anions such as phosphate. At acidic pHs (< pH 7), as prevails in most of the tested wells, and under oxidising conditions the ferric oxides and hydroxides have a resultant net positive charge and therefore have a higher affinity for the negatively charged arsenate oxyanion (Dzombak and Morel, 1990).

For shallow wells (< 40 m), Bowells (1992) studying the weathered profile in the Obuasi area, found that ferric hydroxide and arsenate minerals are relatively abundant products of the arsenopyrite oxidation. In addition, Bowell (1992) noted that under oxidizing and acidic conditions, the ferric arsenates gets destabilized and re-precipitate as haematite ($Fe_2O_3$) and limonite ($FeOOH.nH_2O$) thereby losing the As to the aqueous medium. For wells with shallow depths, assuming presence of sufficient oxygen for arsenopyrite and ferrous oxidation, the levels of dissolved iron is therefore expected to be low.

In deeper aquifers (>40 m), as the sulphide oxidation proceeds, dissolved oxygen gets gradually consumed, the level of ferrous ions and arsenic oxyanions will increase but the subsequent oxidation of the ferrous to ferric oxides will be limited by the low dissolved oxygen. In addition the onset of reducing conditions, could trigger the reduction of Fe(III) and Mn(IV) from existing iron and manganese containing minerals (Bowell, 1992 and Smedley, 1996). Also such reducing conditions could cause sulphate reduction and the release of adsorbed arsenic.

All the wells sampled have a depth in the range of 40 – 60 m and about 70% of the arsenic contaminated wells were found to be acidic (i.e. pH 5.6-6.5). Depending upon the type of soil and the kind of activities occurring within it moderately reducing conditions may prevail in wells of such depth (Smedley, 1996). All the communities namely Ntoboroso, Anyinase, Dumasi, Achichire (also known as Akykyere) and Aserewadi which had arsenic levels >50 μg/l also had high iron content (>1.00 mg/l). Sampled wells in four out of five arsenic contaminated communities had in addition high manganese content (> 0.4 mg/l).

The individual correlation plots of arsenic levels (specifically for wells with [As] > 10 μg/l) versus the dissolved manganese, iron content, pH or age of the arsenic contaminated wells show a very low degree of correlation (Figure 2.7). However a linear multiple regression analysis performed using a 5% level of significance with the arsenic concentration as the dependent variable and the Fe, Mn, pH (i.e. pH < 7), and age of well (i.e. > 15 years) as independent predictors put together, gave some degree of correlation with a multiple $r^2$ value of 0.6963. In the linear multiple regression analysis the null hypothesis employed was the presumption that there is no association between the arsenic concentration and any of the predictors. In this case the rejection criterion for the null hypothesis was that, the obtained p-values should be less than 0.05. From the analysis, the following p-values were obtained for the individual predictors: Fe → 0.01371, Mn → 0.007572, pH → 0.023167, 'age of well' → 0.688575. These results imply that there is significant correlation between arsenic concentration and acidic pH, Fe and Mn contents of the contaminated wells. However no significant correlation exists between arsenic concentration and the age of the contaminated wells.

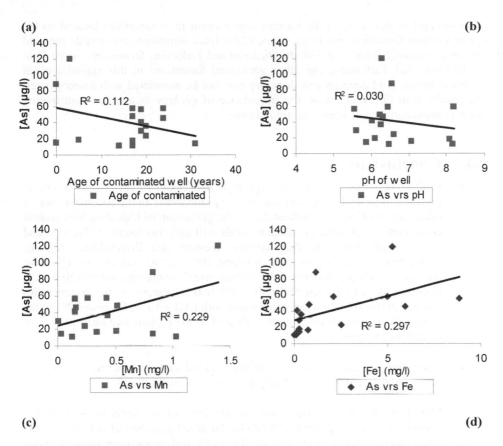

**Figure 2.7** Correlations between arsenic levels versus (**a**) Age of well (**b**) pH (**c**) [Mn] and (**d**) [Fe] of contaminated wells (i.e. [As] >10 μg/l) in Atwima and Wassa Amenfi districts.

From the results, more than 80% of the samples taken from the goldbelt zones did not show any detectable or elevated arsenic contamination (i.e. [As] > 10 μg/l). This trend of occassional arsenic occurrence could be attributed to the complex nature of the geology in the goldbelt zone. The Figures 2.1 and 2.6 depict different formations occurring alongside and within the Birimian and Takwaian formations in the study area that may not harbor arsenopyrite. The wells in these different formations may have no arsenic. Analysis on the detailed geology of the wells was however, beyond this study. Moreover, the mineral composition of the Birimian and Tarkwaian is not homogenous and therefore arsenopyrites may be absent in some areas and present in others within the same formation.

Considering the detailed geology of the Eastern region (Figures 2.1 and 2.6) it is observed that certain parts lie within the Birimian formation whiles other areas are located within the Upper Voltaian, Togo series, Obosum and Oti beds systems. The wells sampled in this study in the Eastern region occur in communities located on the Upper Voltaian formation and Togo series. These latter formations are outside the gold belt zones normally associated with the Birimian and Tarkwaian formations. Apart from the Birimian and Tarkwaian, the other geological formations in this region are not known to harbor gold deposits and therefore may not be associated with arsenopyrites. The results from this study show the importance of geology as a major contributing factor to the occurrence of arsenic in groundwater.

## 2.4    Conclusions

- Unlike other areas in the world e.g. Bangladesh, Southern Taiwan etc. where As presence in groundwater is prevalent, the groundwater in the basement rock of Ghana are relatively less vulnerable to the generation of high dissolved arsenic concentrations. Arsenic presence at levels >10 µg/l was found in 7%, 12% and 5% of wells tested in the Ashanti, Western and Brong-Ahafo regions, respectively. Of all the samples analyzed, the Anyinase community (Western region) groundwater possessed the highest arsenic contamination of 120 µg/l. It is estimated that between 500 000 – 800 000 inhabitants in the communities covered in this study use untreated water with [As] > 10 µg/l. About 70% of the wells with arsenic concentration > 10 µg/l in the regions had been in use for more than 15 years.

- Thirty five percent, 50% and 5% of the sampled wells in Ashanti, Western and Brong -Ahafo regions, respectively had iron content >0.3 mg/l.

- Manganese levels in thirteen percent and 29% of the wells in Ashanti and Western regions exceeded the WHO health-based guideline of 0.4 mg/l. In the Brong-Ahafo region, only 6% of the wells had manganese concentrations beyond the guideline value.

- The communities with high arsenic contamination are situated on the Birimian sediments, Birimian volcanics and Tarkwaian formation in the gold-belt zone.

- All the communities with high arsenic contamination (>50 µg/l) have iron levels >1.0 mg/l and manganese >0.4 mg/l. Most of them have pH within the range of 5.6 to 6.5. The release of arsenic into the aquifers is presumably due to arsenopyrite oxidation and reductive dissolution of ferric oxides and desorption of adsorbed arsenic. A linear multiple regression showed significant correlation between the concentrations of arsenic in well water with acidic pH, Fe and Mn content. There is however no significant correlation between arsenic concentration and the age of the contaminated well.

- Most of sampled wells in the Eastern region, outside the gold-belt zone had iron >0.3 mg/l and manganese >0.4 mg/l but no arsenic contamination. The wells sampled in the Eastern region are located within the Upper Voltain and Togo series geological formations that have no gold deposits. These formations have different characteristic features from the Birimian and Tarkwain formations that are often associated with arsenopyrites. This shows that geology has major impact on the occurrence of arsenic in groundwater.

It is recommended that a country wide inventory of the water quality especially the arsenic occurrence in existing wells should be made and further investigations carried out to establish a possible link between arsenic occurrence and the health status of consumers. In addition a more detailed study on the arsenic mobilization within the depth profile of the aquifers in the arsenic contaminated areas should be carried out.

## 2.5    References

Appelo, C. A. J. and Postma, D.: 1994, Iron in groundwater. In 'Geochemistry, groundwater and pollution.' A. A. A. Balkena Publishers, Rotterdam, Netherlands. pp 279 – 284.

Bowell, R. J. 1992, Supergene gold mineralogy at Ashanti, Ghana: implications for the supergene behaviour of gold. *Mineralogical Magazine* 56; 545 – 560.

Chien-Jen, C.; Lin-I, H.; Chin-Hsiao, T.; Yu-Mei, H. and Hung-Yi, C.,: 1999, Emerging epidemics of arseniasis in Asia. In 'Arsenic Exposure and Health Effects'. Editors: W.R. Chappell, C. O. Abernathy and R. L. Calderon. Publishers: Elsevier.

Junner, H.R. Hirst, T.: 1946, Geology and hydrology of Voltain Basin. Gold Coast Geological Survey Memo. 1946.

Kesse, G.O.:1985, The mineral and rock resources of Ghana. pp 610 Rotterdam, A.A. Balkema.

Leube, A. Hirdes, W., Mauer, R. and Kesse, G. O. 1990, The early proterozoic Birimian supergroup of Ghana and some aspects of its associated gold mineralization. *Precambrian Research* 46:139 – 165.

Mining Portal of Ghana. 2006 http://www.ghana-mining.org. (Last accessed – June 2008)

Nishimura, T., Ito,C.T. and Tozawa, K.S. 1985 The calcium-arsenic-water-air system. Proc. CIM Annual Hydromet. Meeting, Montreal Quebec.

Ntiamoah-Agyakwa, Y., 1977, Gold mineralization in the Precambrian of Ghana. Utilisation Min. Res. Develop. Countries Conf. Proc. Vol.2, Lusaka, pp 1 – 18.

Nkyerkyer, K. 2000, Pattern of gynaecological cancers in Ghana. *East Africa Medical Journal*, vol. 77, No.10: 534 – 538.

Norman, I.D. and Miller, G. P.: 1999, Report on the Arsenic and Water Chemistry Study of    SW Ghana Wells, June – 1999. (done as part of the 1999 Ghana Fulbright-Hays Program-Ghanaian Environmental Science Seminar and Cross-Cultural Studies).

Post, E. P.: 1999, Manganese oxide minerals: Crystal structures and economic and environmental significance. *Proc. Natl. Acad. Sci. USA*. Vol. **96**, pp 3447 – 3454. Colloquium paper.

Siripitayakunkit, U.; Visudhiphan, P.; Pradipasen, M. and Vorapongsathron, T. 1999 Association between chronic arsenic exposure and children's intelligence in Thailand. In 'Arsenic Exposure and Health Effects'. Editors: W.R. Chappell, C. O. Abernathy and R. L. Calderon. Publishers: Elsevier 1999. p141.

Sracek, A., Bhattacharya, P., Jacks, G. and Gustafsson, J.P.: 2001, Mobility of arsenic and geochemical modeling in groundwater environment. In: Groundwater Arsenic Contamination in the Bengal Delta Plain of Bangladesh. Proceedings of the KTH-Dhaka University Seminar. Jacks, G., Bhattacharya, P and Khan A.A.(eds).

Smedley, P.L.: 1996, Arsenic in rural groundwater in Ghana. Journal of African Earth Sciences, Vol. 22, No. 4: 459 – 470.

Smedley, P.L., Edmunds, W.M. and Pelig-Ba, K.B.: 1996, Mobility of arsenic in groundwater in the Obuasi gold mining area of Ghana: some implications for human health. In:

Environmental Geochemistry and Health. Appleton, J.D., Fuge, R. and McCall, G.J.H.(eds). Geological Society Special Publication, pp113, 163 – 181.

Sestini, G.; 1973, Sedimentology of a paleoplacer: the gold bearing Tarkwain of Ghana. In: G.C. Amstutz and A.J. Bernard (eds), Ores in Sediments. Springer, Berlin, pp. 275 – 306.

WHO: 1996, Guidelines for drinking-water quality, 2nd edition, Vol 2. Health criteria and other supporting information. Geneva, World Health Organization.

WHO: 2004, Guidelines for drinking-water quality, 3rd edition, Vol. Geneva, World Health Organization. pp 186, 398.

WHO, (2006) Guidelines for drinking – water quality. First Addendum to Third edition. Vol. 1 recommendations 186 & 398, Geneva
http://www.who.int/water_sanitation_health/dwq/gdwq0506.pdf

Yang, R.T. 1999 'Gas Separation By Adsorption Processes'. – Series on chemical engineering. Vol 1. Publishers: - Imperial College Press.

# CHAPTER THREE

# ADSORPTIVE REMOVAL OF MANGANESE (II) FROM THE AQUEOUS PHASE USING IRON OXIDE COATED SAND

This Chapter was published as:
Buamah, R. Petrusevski, B. and Schippers, J.C. (2008) Adsorptive removal of manganese (II) from the aqueous phase using iron oxide coated sand. *Journal of Water Supply: Research and Technology* – AQUA 57.1:1 – 12.

## Abstract

IOCS (iron oxide coated sand) is a potential adsorbent for metallic ions e.g. arsenic, lead, manganese etc. The effect of process conditions on Mn (II) adsorption on IOCS and related process mechanisms haven't been thoroughly investigated. This study determined the capacity, rate, mechanisms involved and the effect of process conditions on the adsorption of Mn (II) onto IOCS using laboratory scale batch reactors with modeled water.

Alkalinity and pH markedly affected the solubility of Mn (II) that is governed by manganese carbonate; solubility is very limited ($\leq$ 2 mg/l) even at low alkalinity (60 mg/l). The IOCS demonstrated an increasing Mn (II) adsorption capacity ('K' values: 4.7 – 147) with pH increase from 6 to 8. Comparable adsorption capacities were found at pH 6 under both oxic and anoxic conditions for experimental periods of up to 24 hours. This indicates that no significant quantities of adsorbed Mn (II) were oxidized at pH 6 within 24 hours to form extra capacity.

Kinetic studies using the Linear Driving force, Lagergren and Potential Driving Second Order Kinetic (PDSOK) models revealed that the rate of manganese (II) adsorption onto aggregate IOCS declines after the initial phase likely due to the saturation of easily accessible adsorption sites on grain surface and / or pH drop in the pores of the IOCS grains due to Mn (II) adsorption. The changing adsorption rate constants prevented the equilibrium concentration being predicted with the applied models.

**Key words:** adorption, iron-oxide coated sand, manganese (II)

## 3.1    Introduction

Groundwater, by virtue of its generally good and constant quality, is popular as a drinking water resource. The presence of manganese in concentrations exceeding 0.1 mg/L may give rise to complaints about taste, staining of plumbing fixtures and turbidity (WHO 2004). In distribution systems manganese concentrations as low as 0.02 mg/l tend to form coatings on piping which later slough off as black precipitate (Sly et al. 1990; MWH, 2005). The present WHO guideline value of 0.4 mg/l recommended for manganese in drinking water is health-related (WHO, 2004). Exposure to high manganese (> 0.5 mg/l) over the course of years has been associated with toxicity to the nervous system producing a syndrome that resembles Parkinsonism (Fact Sheet 2001).

Generally, aquifers possessing high levels of iron have appreciably high levels of manganese. Manganese concentrations are usually five to twenty times lower than iron concentrations in groundwater. Water treatment plants treating groundwater with high iron and manganese frequently incorporate two filter stages; the first for the removal of mainly iron and some manganese and the second as a polishing step for iron and the removal of the main part of manganese. Sand filters used in the treatment of groundwater with high iron levels may normally end up having coated layers of iron oxides with manganese oxides embedded in the coating. Both the iron and manganese oxides in the coating of the media of such sand filters have a high potential for the adsorption of heavy metal ions e.g. As, Cr, etc (Petrusevski *et al.* 2002). Iron-oxide coated sand (IOCS) has given good indication of being a potential adsorbent for removal of Mn. The iron oxide present within the coating of the IOCS possibly enhances the autocatalytic oxidation of adsorbed $Mn^{2+}$ and thereby forms manganese oxides (Junta & Hochella 1994). However, the mechanisms involved in the adsorption of manganese ions onto the IOCS and particularly the effects of oxic and anoxic conditions have not been well studied. This study seeks to investigate:

- The effect of alkalinity and pH on manganese solubility.
- The adsorption capacity of IOCS for manganese at different pH values.
- The effect of oxic and anoxic conditions on the manganese (II) adsorption.
- The kinetics and mechanism involved in the adsorption process.
- The effect of adsorbed Mn (II) on the rate of oxidation of manganese in the coating of the IOCS under oxic conditions.

## 3.2    Theoretical background

### 3.2.1 Solubility and oxidation states of manganese

Among the numerous oxidation states of manganese, the II, III and IV oxidation states are of greatest environmental importance. Mn (II) in aqueous solutions is stable at pH values below 9 because of its very low rate of oxidation. Therefore, in the range pH 6 to 9 of most natural waters, Mn (II) if present remains stable. Mn (II) oxidation results in the formation of Mn oxides which occur as coatings on soil and sediment particles and as discrete particles (Ehrlich 1996). Mn (II) exists in a variety of minerals such as birnessite $(Na_4Mn_{14}O_{27}.9H_2O)$, hollandite $[Ba(Mn^{4+}.Mn^{2+})_8O_{16}]$, rhodochrosite $(MnCO_3)$, alabanite (MnS), reddingite $[Mn_3(PO_4)_2.3H_2O]$, etc. Mn (III) is thermodynamically unstable in aqueous solutions, being easily reduced to Mn (II) and can even undergo disproportionation reaction (e.g. Equation 3.5) in the absence of reducing agents (Cotton & Wilkinson 1967). Mn (III) does not occur in soluble form except in the presence of strong complexing agents such as humic acids or other organic acids (Kostka *et al.* 1995). Mn (III) and Mn (IV) primarily form insoluble oxides and oxyhydroxides. The solubility of manganese (IV) oxide $(MnO_2)$ is very low within the pH range 3 to 10 (the solubility product: $10^{-41}$, Stumn & Morgan 1996).

The concentration of dissolved Mn in groundwaters and surface waters is largely controlled by redox reactions between Mn (II) and Mn (III, IV) and governed by pH. Mn (II) oxidation in such natural systems is thermodynamically favourable but often proceeds at very slow rates. According to Murray *et al.* (1985), under more extreme conditions (pH > 8.5, oxygen partial pressure of 1 atm or $[Mn_{Total}]$ > 25 mg/L) Mn (II) homogeneously oxidizes within a few weeks to months. The Mn (II) oxidation is autocatalytic with the Mn-(oxyhydr)oxide products adsorbing Mn (II) and catalysing its oxidation (Stumm & Morgan 1996). The Mn (II) oxidation may also be catalysed by a variety of other surfaces including Fe oxides and silicates (Junta & Hochella 1994). Mn (II) oxidation proceeds via a number of reactions yielding various oxidation states and mineral forms (Equations 3.1 to 3.5). Mn (II) can be oxidized directly to Mn (IV) or to mixed Mn (II) and Mn (III) oxides and oxyhydroxides or to $Mn_3O_4$ (composed of $Mn_2O_3.MnO$). These mixed oxides may undergo protonation or disproportionation reactions (-the rate determining step) to form Mn (IV) oxides and manganates i.e. the '6-valent manganese' (Murray et al. 1985).

$$Mn^{2+} + \frac{1}{2} O_2 + H_2O \rightarrow MnO_2 + 2 H^+ \tag{3.1}$$
$$Mn^{2+} + \frac{1}{4} O_2 + 3/2 H_2O \rightarrow MnOOH + 2 H^+ \tag{3.2}$$
$$3 Mn^{2+} + \frac{1}{2} O_2 + 3H_2O \rightarrow Mn_3O_4 + 6 H^+ \tag{3.3}$$
$$Mn_3O_4 + 2 H^+ \rightarrow 2 MnOOH + Mn^{2+} \tag{3.4}$$
$$Mn_3O_4 + 4 H^+ \rightarrow MnO_2 + 2 Mn^{2+} + 2 H_2O \tag{3.5}$$

## 3.2.2 The adsorption phenomenon

The adsorption phenomenon involves the separation of a substance from one phase accompanied by its concentration on the surface of another. The adsorbing phase is the adsorbent and the material concentrated is the adsorbate. The character of the quantitative equilibrium distribution between the phases which affects the adsorption is influenced by a variety of factors. These factors relate to the properties of the adsorbate, the adsorbent and the system in which the adsorption occurs (Slejko 1985).

Adsorption on a surface or interface is the result of binding forces between the individual atoms, ions or molecules of the adsorbate and the surface; these forces have their origin in electromagnetic interactions (Van Vliet & Weber Jr. 1981). The interactions between adsorbate and adsorbent consist of molecular forces embracing permanent dipole, induced dipole and quadruple electrostatic effects. These forces may be classified into short range forces (i.e. the chemical forces) and long range forces (i.e. the coulombic forces). The short range forces may give rise to covalent or hydrophobic bonds or hydrogen bonding or steric effect. The long range forces give rise to electrostatic attraction (Yang 1999). Based upon these adhesive forces four principal types of adsorption have been identified: namely, ion exchange, chemical adsorption, physical adsorption and specific adsorption (Yang 1999).

### 3.2.2.1 Ion exchange

Ion exchange involves short-range forces and normally occurs within materials with porous lattice containing fixed charges. In ion exchange, the electrostatic attachment of ionic species to sites of opposite charge at the surface of an adsorbent occurs with a subsequent displacement of these species by other ionic adsorbates of greater electrostatic affinity. This substitution generally results in the emergence of a net charge on the surface of the adsorbent. Consequently in aqueous solutions the electrostatic attraction brings dissolved ions of opposite charge to the adsorbent to balance the charge (Deutsch 1997). Ion exchange may involve either a cation exchange or an anion exchange and in both cases pH changes play a vital role in the exchange capacity.

### 3.2.2.2 Chemical adsorption

Chemical adsorption involves a reaction between an adsorbate and adsorbent resulting in a change in the chemical form of the adsorbate. Sharing of electrons between the adsorbent and the adsorbate occurs at the adsorptive site of the adsorbent yielding process irreversibility. Adsorption onto hydrous metal oxides surface e.g. hydrous iron oxide or manganese oxide is predominantly, chemical adsorption.

According to Davis and Leckie (1978) the adsorption on hydrous metal oxide involves surface ionization and surface complexation of metal ions with the hydrous metal oxide. In aqueous systems the surface of the metal oxide gets covered with hydroxyl groups. An acid-base equilibrium involving the hydroxylated oxide surface is established as follows:

$$\equiv S\text{-}OH + H^+ \leftrightarrow \equiv S\text{-}OH_2^+ \tag{3.6}$$
$$\equiv S\text{-}OH \leftrightarrow \equiv S\text{-}O^- + H^+ \tag{3.7}$$

where $\equiv S\text{-}OH_2^+$, $\equiv S\text{-}OH$ and $\equiv S\text{-}O^-$ represent positively, neutral and negatively charged surface hydroxyl respectively. From these representations the point of zero charge ($P_{zc}$) can be depicted as '$\equiv S\text{-}OH$'. The $P_{zc}$ is the pH of the solution in chemical equilibrium with the surface in its neutralized state through adsorption of $H^+$ and / or $OH^-$ ions (Casamassima & Darque-certti 1993). Mn oxides have a very negative charge at pH values higher than their $P_{zc}$ and their cation adsorption capacity therefore generally increases with increasing pH (Liu et al. 2004). At pH values below their $P_{zc}$ the net surface charge is positive and anion adsorption is favoured.

In the surface complexation, initially a bond is formed between the metal ion to be adsorbed and the surface oxygen atom of the hydrous metal oxide resulting in the release of protons and a drop in pH:

$$\equiv S\text{-}OH + M^{2+} \leftrightarrow \equiv S\text{-}OM^+ + H^+ \tag{3.8}$$

where $M^{2+}$ is a divalent cation.

In aqueous solution with low pH values, the surface is more positively charged because of the additional complexed hydrogen ions producing $\equiv S\text{-}OMOH_2^+$ and loss of $OH^-$.

$$\equiv S\text{-}OH + M^{2+} + H_2O \leftrightarrow \equiv S\text{-}OMOH_2^+ + H^+/OH^- \tag{3.9}$$

At a higher pH condition the surface is predominantly negatively charged due to the loss of $H^+$ from the surface and subsequently the surface becomes more attractive to cations.

$$\equiv S\text{-}OH + M^{2+} + H_2O \leftrightarrow \equiv S\text{-}OMO^- + 3H^+ \tag{3.10}$$

Thus adsorption of cations on to hydrous oxides increases with increase in pH. However, according to Dzombak and Morel (1990) there exist a narrow pH range (usually one or two units) within which the percentage adsorption of cations onto hydrous oxides increases from 0 to 100% giving typical adsorption 'pH edges'.

Anion adsorption by hydrous oxides occurs by a different mechanism.   Anions are adsorbed via a ligand exchange reaction in which hydroxyl surface groups are replaced by the sorbing ions as shown in the following reactions:

$$\equiv S\text{–}OH + A^{2-} + H^{+} \leftrightarrow \equiv SA^{-} + H_2O \qquad (3.11)$$
$$\equiv S\text{–}OH + A^{2-} + 2H^{+} \leftrightarrow \equiv SHA + H_2O \qquad (3.12)$$

where $A^{2-}$ is the hypothetical divalent anion.

### 3.2.2.3 Physical adsorption

Physical adsorption generally results from the action of van der Waals forces that hold the adsorbate molecule to the atoms on the adsorbent surface. The adsorbed molecule is not affixed to a specific site but free to undergo translational movements.

### 3.2.2.4 Specific adsorption

Specific adsorption involves attachment of adsorbate molecules at functional groups on adsorbent surfaces. In these interactions the adsorbate does not undergo any transformation. Specific adsorption exhibits a range of binding energies common to both chemical and physical adsorption.

## 3.2.3 Freundlich's adsorption isotherm

Freundlich's Adsorption isotherm is commonly used to quantify the adsorption capacity of different adsorbents; however, the Langmuir isotherm can be used as well. In this article, the Freundlich's Isotherm is used. Expressed mathematically:

$$q_s = Kc_s^{1/n} \qquad (3.13)$$

where:
$q_s$ = amount of adsorbate adsorbed per unit mass or surface area of the adsorbent (g/g or g/m$^2$)
$c_s$ = equilibrium concentration of the adsorbate (g/m$^3$).
K and n = isotherm constants.

K [mg/g x (mg/l)$^n$] gives an indication of the adsorption capacity while n (a constant with no dimension) is a measure of the adsorption intensity and also reflects the steepness of the curve whether plotted on an arithmetic or logarithmic scale (Faust & Aly 1998).

## 3.2.4 Adsorption kinetic models

Several kinetic models are available for studying adsorption kinetics. In this article three models, namely the Linear Driving Force model (LDF), the Lagergren model and the Potential Driving Second Order Kinetic model (PDSOK) are applied to investigate the rate of adsorption of Mn (II) onto IOCS in batch experiments.

### 3.2.4.1 The Linear Driving Force model

The Linear Driving Force Model (LDF) is an approach which employs an overall kinetic rate constant (covering both internal and external mass transfer coefficients) in defining the adsorption rate equation. According to the LDF model equation, the rate of adsorption of an adsorbate (i.e. Mn (II)) by an adsorbent (in this case IOCS) is linearly proportional to the driving force. The driving force is defined as the difference between the surface concentration and the average adsorbed phase concentration (Tien 1994).

Mathematically,

$$dq/dt = k_1 (q_s - q) \qquad (3.14)$$

where:
$q$ = the solid phase manganese (II) concentration (kg/kg)
$q_s$ = the solid phase manganese (II) concentration in equilibrium with $c_s$ (kg/kg)
$k_1$ = the LDF kinetic rate constant ($s^{-1}$)
$t$ = time (s)

Using Freundlich's isotherm equation and mass balance, the above equation could be re-written expressing the q in terms of the initial liquid phase manganese (II) concentration:

$$c = (c_o - c_s)e^{-k_1 t} + c_s \qquad (3.15)$$

In natural logarithimic form:

$$\ln [(c - c_s) / (c_o - c_s)] = -k_1 t \qquad (3.16)$$

where:
$c$ = the liquid phase manganese (II) concentration ($kg/m^3$) at time t
$c_o$ = the initial liquid phase manganese (II) concentration ($kg/m^3$) at time $t_o$.
$c_s$ = manganese (II) concentration ($kg/m^3$) in the liquid phase at equilibrium

$k_1$, the LDF rate constant, can be obtained from the kinetic data of a batch adsorption experiment.

### 3.2.4.2 The Lagergren model

To determine the order of the adsorption, the Lagergren's equation can be applied (El-Zawahry & Kamel, 2004).

$$\log (q_s - q) = \log q_s - t k_2 / 2.303 \qquad \text{(Lagergren's Equation)} \qquad (3.17)$$

where:
$q$ = the amount of $Mn^{2+}$ (g/l) adsorbed at time t.
$q_s$ = the amount of $Mn^{2+}$ adsorbed (g/l) at equilibrium

$k_2$ = the rate constant for the adsorption process; $k_2$ can be determined from the slope of the linear plot of log $(q_s - q)$ versus t. Linearity of plot will be an indication that the reaction taking place is of the first order.

### 3.2.4.3 The Potential Driving Second Order Kinetic model

The PDSOK assumes that the most important factors influencing the adsorption process are the chemical potentials of both the adsorbent surface and the solution. The chemical potentials are, in turn influenced by the contact time, temperature, pH, adsorbent concentration, nature of the solute and its concentration. The PDSOK model could be used to investigate among others, the order of the reaction (Liu *et al.* 2004). For the PDSOK model, the rate differential equation for the adsorption process is expressed as follows:

$$dc/dt = - k_3(s_s - s)(c - c_s) \qquad (3.18)$$

where:
$c_s$ = $Mn^{2+}$ concentration (mol/l) in the liquid phase at equilibrium
$c$ = $Mn^{2+}$ concentration (mol/l) in the liquid phase at time t
$s$ = the number of active sites on the adsorbent occupied (mol/g) at time t
$s_s$ = the number of active sites on the adsorbent occupied (mol/g) at equilibrium
$a$ = the adsorbent dosage (g/l)
$k_3$ = the rate constant

This implies that:

$$s = (c_o - c) / a \qquad (3.19)$$

where:
$c_o$ = $Mn^{2+}$ concentration (mol/ l) in the liquid phase at time $t_o$

$$s_s = (c_o - c_s) / a \qquad (3.20)$$

Substituting for s and $s_s$ in Equation (3.18) and rearranging gives:

$$- dc / (c_s - c)^2 = k_3 dt /a \qquad (3.21)$$

Integrating with boundary conditions t = 0 and t = t and $c_o$ to c yields:

$$1 / [(c_o - c_s)-(c_o - c)] = 1 / (c_o - c_s) + k_3 t / a \qquad (3.22)$$

Rearranging Equation (3.22) gives rise to the following:

$$t / (c_o - c) = t / (c_o - c_s) + a / k_3(c_o - c_s)^2 \qquad (3.23)$$

For a unit mass of adsorbent, the following is obtained:

$$t / (c_o - c) = t / (c_o - c_s) + 1 / k_3(c_o - c_s)^2 \qquad (3.24)$$

The rate constant $k_3$ and equilibrium concentration $c_s$ can be determined experimentally by plotting $t / (c_o - c)$ against t.

## 3.3    Materials and Methods

### 3.3.1 Materials for batch studies

In this study, IOCS collected from the Noord Bargeres groundwater treatment plant in the Netherlands, was used as the adsorbent. The main function of rapid sand filters at the plant is the removal of iron, manganese and some ammonium. The water quality of the groundwater treated at the plant is given in Table 3.1 (Sharma 2002).

**Table 3.1** Composition of groundwater treated at the treatment plant in Noord Bargeres.

| pH | Fe (mg/l) | Mn (mg/l) | $Ca^{2+}$ (mg/l) | $Mg^{2+}$ (mg/l) | Cl⁻ (mg/l) | $SO_4^{2-}$ (mg/l) | $HCO_3^-$ (mg/l) | $NH_4^+$ (mg/l) | DOC (mg/l) |
|---|---|---|---|---|---|---|---|---|---|
| 6.9 | 13.3 | 0.49 | 50 | 6.5 | 36 | 57 | 128 | 0.2 | 2.0 |

From the composition analysis on the IOCS, it was found that 509 mg iron and 24.5 mg manganese were present per gram of IOCS. Both pulverised form (particle size: < 65 μm) and aggregate form (particle size: 2.00 – 3.15 mm) of the IOCS were used in the study. The pulverised IOCS was used in order to provide a larger surface area which would increase the rate of adsorption. The pulverised IOCS was used in most experiments since the adsorption process on aggregate IOCS was very slow and predictions of the ultimate adsorption capacity by making use of kinetic models were not successful. At most water treatments plants, the sand filter media normally remains in use for several years before replacement. During the period, the adsorbed ions penetrate and diffuse into the pores within the coating of the sand grains thereby gradually increasing the adsorption capacity of the coated sand grain with time. The use of aggregate media for batch experiments may underestimate the adsorption potential of the media. With the pulverised media, much of the less accessible pores are opened up and the adsorption capacities determined from the experiment would be close to the ultimate adsorption capacity of the media. Over-estimation of the capacity may however, not be excluded in this case.

To ascertain the role of virgin sand material in the adsorption experiments, de-coated IOCS was tested for its manganese adsorption capacity. For this purpose 1 g of pulverised IOCS was treated in 6N HCl solution over 24 hours to dissolve the coating completely, washed with de-mineralised water and dried.

#### 3.3.1.1 Batch experiments

Batch experiments were done under oxic and anoxic conditions with model water. Most groundwaters have Mn (II) concentrations of up to 2 mg/L however on rare occasions

Mn (II) concentrations could go up to about 10 mg/L. A stock solution of concentration 1000 mg Mn (II) /L was prepared using analytical-grade $MnSO_4.H_2O$ and acidified with concentrated HCl (32% assay) to a pH < 2. The maximum Mn(II) concentrations at the various pH for which interference of possible $MnCO_3$ precipitation is avoided, were calculated using solubility product equations (Stumn & Morgan 1996) and the Phreeq C software programmes (Parkhurst & Appelo 1999). Experiments were performed with initial Mn (II) concentrations of up to 10 mg/L to enable a high accuracy, prevent manganese carbonate precipitation and simulate the range of manganese levels in groundwaters. To ensure a stable pH during experiments, $NaHCO_3$ was added to the model water.

For the oxic experiments acid-cleansed 500 ml plastic bottles fitted with tubes for periodic sampling were filled with de-mineralized water dosed with the appropriate amount of $NaHCO_3$ (Table 3.2) and their pH subsequently adjusted with 6N HCl or 1N NaOH solutions to the required level. After thorough mixing of the prepared model water, Mn (II) was dosed and 4 g of pulverised, aggregate IOCS or de-coated IOCS were added. Bottles were kept at 20 ± 1 °C and placed on a shaker operated at 80 rpm. Blank tests were carried out without the addition of (de-coated) IOCS.

Experiments under anoxic conditions were done in 1.5 L glass reactors with model water (Table 3.2) and a dosage of 8 g IOCS/ l (Figure 3.1). Nitrogen gas was infused into the reactor to attain and maintain anaerobic conditions. Carbon dioxide was introduced for pH adjustment. Mixing was ensured by the gas infusions and continuous stirring. Periodic sampling was done at regular time intervals to determine the rate of adsorption and equilibrium conditions. Equilibrium was considered to have been reached when the difference in manganese concentration of two consecutive samples taken over a period of 10 hrs was ≤ 0.02 mg/l. All experiments were run in duplicate and samples filtered through a 0.45 µm membrane filter using a polypropylene syringe filter. Manganese analysis was done using Perkin Elmer 3110 spectrometer in accordance with the Dutch Standard Method NEN 6457. The method has a detection level of 0.02 mg Mn (II)/ l.

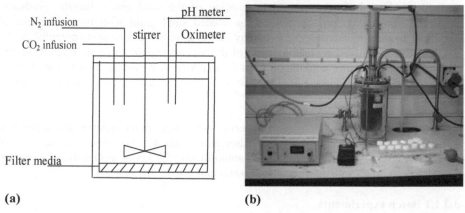

(a)                                          (b)

**Figure 3.1** (a) A schematic diagram and (b) a photograph of the experimental set-up used for anoxic batch adsorption experiments.

**Table 3.2** Model water composition for the batch experiments.

| | Oxic experiment | | | | Anoxic experiment | | | |
|---|---|---|---|---|---|---|---|---|
| | pH5 | pH6 | pH7 | pH8 | pH5 | pH 6 | pH 7 | pH 8 |
| [NaHCO$_3$] (mg/l) | 1000 | 1000 | 84 | 42 | 1000 | 1000 | 84 | 84*** |
| [Mn$^{2+}$] (mg/l) | 2 | 2 | 2 | 2 | 0.1 – 10 | 0.1– 10 | 0.1 – 8 | 0.1–1.0 |

*** In cases where experiments performed at pH 8 required initial manganese concentrations of 2 mg/l, the sodium hydrogen carbonate concentration was 42 mg/l.

The LDF, Lagergren's and PDSOK models were fitted with the data from the batch experiments performed at pH 6 under anoxic conditions.

## 3.4    Results and Discussions

### 3.4.1 The effect of pH and hydrogen carbonate concentration on the solubility of manganese (II)

The calculated data obtained from the solubility product equations and the Phreeqc software programme were used to develop Figure 3.2 which shows the solubility of Mn (II) as a function of pH and hydrogen carbonate concentration. From Figure 3.2 it follows that at pH 7 and 8 and hydrogen carbonate concentration of 61 mg/l, concentration of dissolved manganese in model water is limited to below 10 mg/l and 1 mg/l respectively. The blank tests performed confirmed that upon exceeding

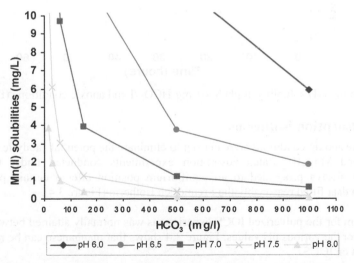

**Figure 3.2** Effect of alkalinity on manganese (II) solubility at various pH values (markers are calculated data).

approximately 1 mg Mn (II)/l concentration in the aqueous solution at pH 8 with 61 mg $HCO_3^-$ /l, precipitation just begins to occur with the passage of time (Figure 3.3). In the batch adsorption experiments conducted in this study, concentrations of manganese and hydrogen carbonate in the model water were selected such as to avoid precipitation of manganese carbonate.

At pH 5 and 6 solubility of manganese carbonate is sufficiently high to avoid precipitation even at the hydrogen carbonate concentration of 1000 mg/l. Groundwater frequently demonstrates high $HCO_3^-$ concentrations of up to 450 mg $HCO_3^-$ /l. These values, together with pH values above 7 might explain why manganese concentrations in groundwaters do not usually exceed 2 mg/l. In groundwater treatment, the standard practical process of aeration prior to filtration normally results in an increase in pH as carbon dioxide is stripped off.   With this practice, the precipitation of manganese carbonate might play a beneficial role in manganese removal especially when the hydrogen carbonate and pH are at relatively high levels.

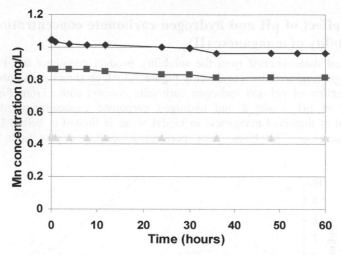

**Figure 3.3** Mn (II) solubility at pH 8, 61 mg $HCO_3^-$/l and anoxic conditions (Blank test).

## 3.4.2 Adsorption isotherms

Initially the anoxic conditions were chosen to eliminate the potential catalytic oxidation of adsorbed Mn (II). Batch adsorption experiments conducted under the anoxic conditions always proceeded to an equilibrium position after a certain period. The adsorption data fitted reasonably the Freundlich isotherm (Figure 3.4).

Equilibrium for the pulverized IOCS experiments was normally attained between 10 and 48 hours for the various pH values. Figure 3.4 shows that manganese can be removed by adsorption on IOCS.

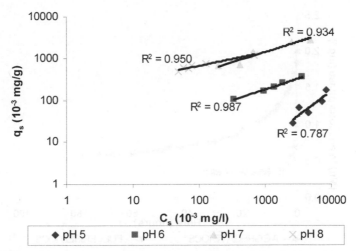

**Figure 3.4** Freundlich isotherms for anoxic manganese adsorption on pulverized IOCS. Model water: 1000 mg $NaHCO_3^-$ /l for pH 5 and 6 and 84 mg $NaHCO_3^-$ /l for pH 7 and 8.

The K values of Freundlich adsorption isotherms established at different pH values (Table 3.3) show pronounced increase in the manganese adsorption capacity of the pulverized IOCS with increasing pH. This increase in the adsorption capacity together with expected high rate of oxidation at higher pHs (Stumn & Morgan 1996) is in consonance with what was observed in practice.

**Table 3.3** The Freundlich's isotherm constants obtained at the various pH values.

|   | pH 5 | pH 6 | pH 7 | pH 8 |
|---|------|------|------|------|
| K | 0.0024 | 4.73 | 45.84 | 147 |
| n | 0.83 | 1.91 | 2.02 | 3.08 |

Unit of K: $[(10^{-3} \text{ mg/g}) / (10^{-3} \text{ mg/l})^{1/n}]$

### 3.4.2.1 Adsorption equilibrium studies
*Comparative performance of pulverized and aggregate IOCS*
Under anoxic conditions at pH 6, it was observed that pulverised IOCS had equilibrium state established after 10 – 12 hours, having achieved about 80% manganese removal at an initial concentration of 2 mg Mn (II)/l and adsorbent dosage of 8 g IOCS/l (Figure 3.5). Equilibrium for the IOCS (aggregate) was established after 80 hours, having removed about 70% of manganese. Pulverised IOCS consequently demonstrated a higher adsorption rate than aggregate IOCS. This difference in capacity is attributed to the opening and shortening of pores as a consequence of grinding. In the aggregate form, a part of the pores is most likely not accessible.

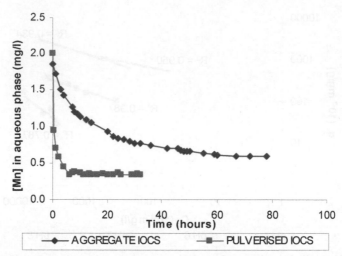

**Figure 3.5** Performance of aggregate and pulverised IOCS in manganese adsorption under anoxic conditions at pH 6. Model water with 1000 mg $NaHCO_3^-$ /l and initial Mn (II) concentration of 2 mg/l and 8 g IOCS /l.

### 3.4.2.2 The effect of oxic and anoxic conditions on the manganese adsorption

The trends of manganese adsorption onto the aggregate media under oxic and anoxic conditions were found to be similar within the first 24 hours of operation (Figure 3.6a). This indicates that no substantial quantities of adsorbed Mn (II) were oxidized under the oxic conditions at pH 6 to form extra capacity within the experimental period of 25 hours.

Under both oxic and anoxic conditions in the reactor, the adsorption process was faster than in the shaker under oxic condition. This indicates that the reactor, using a mechanical stirrer, created a more effective (external) mass transfer (Figure 3.6a).

In Figure 3.6b, the de-coated IOCS demonstrated a very low manganese adsorption capacity thereby confirming the fact that the adsorption process was predominantly due to the adsorptive activity of the mineral coating of the media. A blank test performed in the reactor without IOCS resulted in a negligible reduction of manganese concentration (Figure 3.6b).

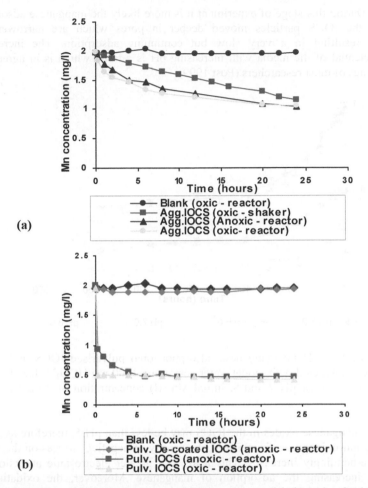

(a)

(b)

**Figure 3.6** Comparative performance of (a) aggregate and (b) pulverised IOCS in Mn (II) adsorption under oxic and anoxic conditions using 8g IOCS /l in both reactor and shaker experiments. Model water: pH6, 1000 mgNaHCO$_3$/l and initial Mn (II) concentration of 4 mg/l.

### 3.4.2.3 Effect of pH on manganese adsorption under oxic conditions

Figure 3.7 demonstrates that adsorption under oxic conditions on pulverised IOCS takes a very long time (250 hours). This effect might be attributed to a catalytic effect of adsorbed Mn (II) that has been oxidised to MnO$_2$ or Mn$_3$O$_4$ (refer Equations 3.1 to 3.3). The latter oxides might have acted as a catalyst and created new adsorption sites, however, the amount of adsorbed manganese is much lower than the amount originally present within the coating. Owing to the lower manganese levels that were present in the model water at the latter stages (i.e >100 hours) for the higher pHs 7 and 8, the expected catalytic effect due to the oxidation of adsorbed manganese could not be ascertained

(Figure 3.7). During this stage of experiment it is more likely the manganese adsorption front within the IOCS particles moved deeper in pores which are narrower and consequently resulting in a very slow but continuing adsorption. The increasing adsorption potential of the media with increasing pH is a trend which is in agreement with the findings of most researchers (Post 1999).

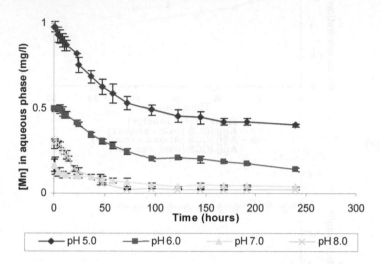

**Figure 3.7** Effect of pH on manganese adsorption onto pulverised IOCS under oxic conditions. Test conducted with model water containing 1000 mg $HCO_3^-$/l, for pH 5 and 6 and 42 mg $HCO_3^-$/l for pH 7 and 8; initial Mn (II) concentration of 2 mg/l and 8 g IOCS/l.

The $P_{zc}$ of the manganese oxides in the IOCS may be less than pH 5, therefore as pH of the aqueous solution increases from 5 to 8, the density of negative charges on the IOCS particles correspondingly increases. This leads to a greater electrostatic attraction for cations, thus increasing the adsorption of manganese. Moreover, the oxidation of adsorbed Mn (II) could also be enhanced with increasing pH. The oxic experiments showed that the manganese removal proceeded with a drop in pH; indicating a release of protons. This evidence signifies that the manganese removal presumably involves the chemical adsorption and / or ion exchange as the predominant removal mechanism.

### 3.4.3 Adsorption kinetic study

Some of the kinetic data obtained from the anoxic experiments with the reactor were tested by fitting to the LDF, Lagergren and PDSOK models (Figure 3.8). For the LDF model (Figure 3.8a) it was found that the slope for the aggregate IOCS was initially steep, subsequently flattened and finally increased again. The pulverised IOCS demonstrated similar behaviour: however, the flattening trend was less distinct. This phenomenon of the pulverised IOCS might be attributed to the faster exhaustion of the adsorption sites which are easily accessible and their pores possibly shorter in length.

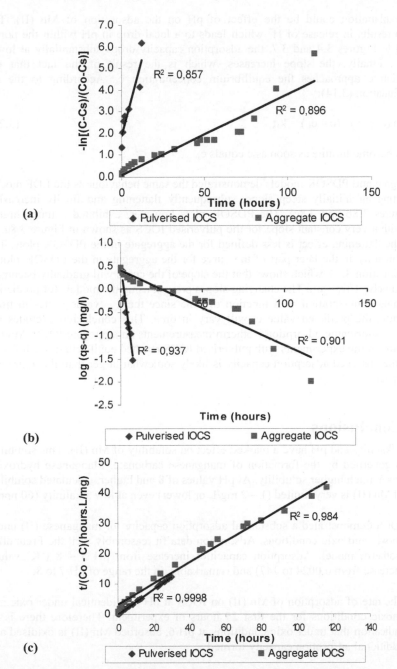

**Figure 3.8** Kinetic models: (a) Linear driving force model (LDF); (b) Lagergren's model and (c) Potential driving second order kinetic model (PDSOK). Test conducted at pH 6 with 1000 mg NaHCO$_3$/l, initial Mn (II) concentration of 4 mg/l and 8 g IOCS/l.

Another explanation could be the effect of pH on the adsorption of Mn (II). The adsorption results in release of $H^+$ which leads to a local drop in pH within the pores. According to Figures 3.4 and 3.7, the adsorption capacity drops substantially at lower pH values. Finally, the slope increases, which is the result of the fact that the concentration c approaches the equilibrium concentration $c_s$. According to the rearranged Equation (3.14):

$$\ln (c_o - c_s) / (c - c_s) = k_1 t \qquad (3.25)$$

$k_1$ tends to become infinite as soon as c equals $c_s$.

The Lagergren and PDSOK models demonstrated the same behaviour as the LDF model; demonstrating an initially steep slope, subsequently flattening and finally increasing again (Figures 3.8b and 3.8c). The PDSOK plots however, exhibited a much higher linearity with a very constant slope for the pulverised IOCS as shown in Figure 3.8c. In addition, the flattening effect is less defined for the aggregate in the PDSOK plots. The observed linearity in the later part of the curve for the aggregate in the PDSOK plot is linked to Equation 3.23, which shows that the slope of the curve will gradually become a constant namely, $1/(c_o - c_s)$. The changing slopes prevent using the models for predicting the remaining concentration as a function of time since the k – value varies in time. Consequently, the predicted value $c_s$ will vary in time. This conclusion excludes the possibility of shortening adsorption isotherm measurements on aggregate IOCS. Mn (II) concentration in the experiments with pulverised IOCS reached equilibrium much faster: however, the observed adsorption capacity is likely somewhat larger than the aggregate IOCS's capacity.

## 3.5    Conclusions

- Alkalinity and pH have a marked effect on solubility of Mn (II). This solubility is governed by the formation of manganese carbonate. Manganese hydroxide has a much higher solubility. At pH values of 8 and higher, calculated solubility of Mn (II) is very limited (1 – 2 mg/L or lower) even at low alkalinity (60 ppm).

- IOCS demonstrated a substantial adsorption capacity for manganese (II) under anoxic and oxic conditions. Adsorption data fit reasonably well the Freundlich isotherm model. Adsorption capacities increase from pH 5 – 8 ('K' values increase from 0.0024 to 147) and remarkably, in the range of pH 7 to 8.

- The rate of adsorption of Mn (II) on IOCS at pH 6 is identical under oxic and anoxic conditions for the first 25 hours of experiment. Therefore there is no indication that under oxic conditions at pH 6, adsorbed Mn (II) is oxidised and additional adsorption capacity formed.

- Plots of the kinetic models, namely Linear driving force, Lagergren and Potential Driving Second Order Kinetic models show a reasonable linearity. However, for aggregate IOCS, the initial slopes decreased with progressing adsorption. This phenomenon might be attributed to the presence of easily accessible and less accessible adsorption sites (e.g. in narrow pores) and / or decreasing pH in the pores as a result of Mn (II) adsorption on IOCS.

- Predicting the equilibrium concentration making use of three kinetic models turned out to be not a feasible option due to changing adsorption rate constants in the course of the adsorption process.

## 3.6    References

Casamassima, M. & Darque-ceretti, E. 1993 Correlation between Lewis donor/ acceptor properties determined by XPS and Bronsted acid/ base properties determined by rest-potential measurements, for aluminium and silicon oxides. *Journal of materials science* **28**, 3997 – 4002.

Cotton, F. A. & Wilkinson, G. 1967 The transition elements. In: Advanced Inorganic Chemistry A Comprehensive Text. 2nd Edition. Interscience Publishers – A division of John Wiley & Sons, New York, pp. 837 – 838.

Davis, J. A.& Leckie, J. O.1978 Surface ionization and complexation at the oxide / water interface: surface properties of amorphous iron oxyhydroxide and adsorption of metal ions. *J. Colloid Interface Sci.* **67** (1), 90 – 107.

Deutsch, W. J. 1997 Water / rock interactions. In: Groundwater Geochemistry – Fundamentals and Applications to Contamination. Lewis Publishers – CRC Press LLC, Boca Raton, Florida – USA. pp. 47 – 60.

Dzombak, D. A. and Morel, F. M. M.: (1990) Surface complexation modelling – Hydrous Ferric Oxides. John Wiley & Sons U. S. A.

Ehrlich, H. L. 1996 *Geomicrobiology*. Marcel Dekker, New York, USA.

El Zawahry, M.M. & Kamel, M.M. 2004 Removal of azo and anthraquinone dyes from aqueous solutions by Eichhornia Crassipes. *Water research* **38**, 2967 – 2972.

Fact Sheet July 2001 *What You Need to Know About Manganese in Drinking Water.* http:www.dph.state.ct.us/publications/BCH/EEOH/manganese Division of Environmental Epidemiology and Occupational Health. Hartford, CT 06134 – 0308.

Faust, S. D. & Aly, O.M. 1998 Chemistry of water treatment: 2nd Edition. Ann Arbor Science book. Lewis Publishers – CRC Press LLC, Boca Raton, London, New York, Washington D.C. pp 145 – 146.

Junta, J. L. & Hochella, Jr. M. F. 1994 Manganese (II) oxidation at mineral surfaces: a microscopic and spectroscopic study. *Geochim Cosmochim Acta* **58**, 4985 – 4999.

Kostka, J. E., Luther III, G. W. & Nealson, K. H. 1995 Chemical and biological reduction of Mn (III) pyrophosphate complexes – potential importance of dissolved Mn (III) as an environmental oxidant. *Geochim Cosmochim Acta* **59**, 885 – 894.

Liu, D., Sansalone, J. J. & Cartledge, F. K. 2004 Adsorption characteristics of oxide coated buoyant media (ℓs < 1.0) for storm water treatment I: Batch Equilibra and Kinetic models. *Journal of Environmental Engineering* **130**(4), 383 – 390.

Murray, J. W., Dillard, J. G., Giovanoli, R., Moers, H. & Stumm, W. 1985 Oxidation of Mn(II): initial mineralogy, oxidation state and aging. *Geochim Cosmochim Acta* **49**, 463 – 470.

MWH (2005) Water Treatment Principles and Design. (2 ed.) Revised by Crittenden, J.C., Trussell, R.R., Hand, D.W., Howe, K.J. and Tchobanoglous, G. John wiley & sons, Hoboken, NJ, USA pp. 1578.

Parkhurst, D.L. & Appelo, C.A.J. 1999 *User's Guide to Phreeqc (Version 2) – A Computer Program for Speciation, Batch-Reaction, One-dimensional Transport and Inverse*

*Geochemical Calculations.* U.S. Geological Survey Water-Resources Investigations Report 99-4259.

Petrusevski, B., Boere, J., Shahidullah, S. M., Sharma, S. K. & Schippers, J. C. 2002 Adsorbent based point-of-use system for arsenic removal in rural areas. *J. Water Supply Res. Technol.* **51**, 135 – 144.

Post, J. E. 1999 Manganese oxide minerals: Crystal structures and economic and environmental significance. Colloquim paper. *Proc. Natl. Sci. USA* **96**, 3447 – 3454.

Sharma, S.K. 2002 *Adsorptive Iron Removal from Groundwater.* PhD Thesis. Unesco-IHE, Institute for Water Education and Wageningen University. Swets and Zeitlinger B.V., Lisse.

Slejko F.L. 1985 Adsorption Technology – A step-by-step Approach to Process Evaluation and Application. Chemical Industries Series Vol. 19. Editor: F.L. Slejko. Tall Oaks Publishing Inc. Voorhees, New Jersey, pp 18 – 30.

Sly, L. I.,Hodgkinson, M.C., and Arunpairojana, V. (1990) Deposition of manganese in a drinking water distribution system. *Appl. Environ. Microbiol.*, 56,(3) 628 – 639.

Stumm, W. & Morgan, J. J. 1996 *Aquatic Chemistry: Chemical equilibria and Rates in Natural Waters.* $3^{rd}$ Edition. Wiley, New York, USA.

Tien, C. 1994 Adsorbate Transport: Its Adsorption and Rates. In: Adsorption Calculations and Modeling. Butterworth – Heinemann series in chemical engineering. Series Editor: Howard Brenner. Publishers - Butterworth – Heinemann, Boston,USA, pp 81 – 82.

Van Vliet, B.M. & Weber Jr. W.J. 1981 Comparative performance of synthetic adsorbents and activated carbon for specific compound removal from wastewaters. *J. Water Poll. Control Fed.* 53 (11), 1585 – 1598.

WHO, (Geneva) 2004 Guidelines for Drinking – Water Quality. $3^{rd}$ Edition. Vol.1 Recommendations. Publishers – SNP Best-set Typesetter Ltd. Hong Kong and Sun Fung – China. pp 397 – 399 & 492.

WHO, (2006) Guidelines for drinking – water quality. First Addendum to Third edition. Vol. 1 recommendations 186 & 398, Geneva
http://www.who.int/water_sanitation_health/dwq/gdwq0506.pdf

Yang, R.T. 1999 *Gas Separation by Adsorption Processes. – Series on Chemical Engineering.* Vol. 1. Publishers – Imperial College Press, London, UK.

# CHAPTER FOUR

# MANGANESE (II) ADSORPTION CHARACTERISTICS OF SELECTED FILTER MEDIA FOR GROUNDWATER TREATMENT: EQUILIBRIUM AND KINETICS

This Chapter was presented at the World Water Congress in Vienna and published as:
Buamah, R. Petrusevski, B. and Schippers, J.C. (2008) Manganese (II) adsorption characteristics of selected filter media for groundwater treatment: equilibrium and kinetics. In: Proceedings of the *IWA Annual Water Congress at Vienna*, 7 – 12 September, 2008 pp 149.

# Abstract

Many ground water treatment plants apply aeration followed by rapid sand filtration for manganese removal. The mechanism involved is assumed to be adsorption of manganese on filter media followed by oxidation. This chapter is mainly focused on the adsorptive part of the process. The manganese adsorption capacities of several filter media were studied under anoxic conditions in batch experiments. The results indicated the following order of manganese adsorption capacities at pH 8: Aquamandix > iron oxide coated sand > iron-ore > manganese green sand > laterite > virgin sand. The experimental equilibrium adsorption data of most of the tested media fit well the Freundlich and Langmuir isotherm equations. Manganese adsorption capacities onto the media were significantly higher at pH 8 as compared with pH 6. The observed adsorption potential of the Aquamandix, iron oxide coated sand (with high and low manganese) and manganese green sand coincided with the manganese contents of these media. This observation suggests that, manganese content in the coating has an important influence on the adsorption capacity of a filter media. The obtained results at pH 8 under oxic conditions indicated auto-catalytic oxidation of adsorbed manganese. Three kinetic models were applied to model the rate of adsorption / removal in batch tests. The Potential Driving Second Order Kinetic model gave the best fit, while the Linear Driving Force Model (Lagergren equation) demonstrated a rather poor fit. The reasonable good fit in the Dubinin-Kaganer-Radushkevisch model together with the high adsorption energies suggest that chemi-sorption is the dominant adsorption mechanism.

**Keywords**: adsorption, equilibrium, isotherm, kinetics, manganese

## 4.1    Introduction

Manganese usually occurs along with iron in the soluble form in most ground waters. Upon exposure to oxygen manganese gets slowly oxidized to the manganese oxides and oxy-hydroxides resulting in undesirable aesthetic problems for users. The presence of manganese in concentrations exceeding 0.1 mg/l may give rise to complaints about taste, staining of plumbing fixtures and turbidity (WHO, 2004). Groundwater, by virtue of its generally good and constant quality, is preferred as a drinking water resource. Many ground waters don't contain dissolved oxygen and in such a situation bivalent iron and manganese exist in dissolved form. Conventionally iron and manganese are often removed through processes like aeration followed by rapid sand filtration or aeration, addition of an oxidant, followed by rapid sand filtration. Several research studies in the past have focused on investigating the removal, oxidation and oxidation kinetics of dissolved manganese in aqueous solutions (Katsoyiannis and Zouboulis, 2003; Po-Yu et al, 2004), however, the adsorption of manganese onto filter media and the oxidation of adsorbed manganese have not been thoroughly and conclusively elucidated.

Manganese and iron oxides have been considered as the principal components controlling the adsorption of heavy metals and metalloids like chromium, arsenic, manganese etc. present in water (Yan-Chu 1994). For example, when manganese or iron oxides are present in filter media used in the treatment of arsenic contaminated water for drinking water production, they tend to oxidize the arsenite ligand (Oscarson et al 1981; Scott and Morgan 1995) and subsequently adsorb the arsenate formed (Takamatsu et al, 1985). Recent studies with the UNESCO-IHE family filters in Bangladesh have shown that IOCS could be very effective in the removal of arsenic and iron from groundwater however if the manganese content of the mineral coating of the IOCS is low, the effectiveness of the latter for manganese removal was found to be limited especially when the groundwater contained high ammonium content (i.e. > 2 mg/L) (Barua, 2006). These studies therefore give reasons to expect that the presence of both manganese oxides and hydrous iron oxides in a filter media are essential for a good manganese adsorption potential. This expectation however, needs to be verified.

In this study the comparative performance of various potential filter media including iron oxide coated sand (IOCS), manganese green sand, laterite, iron ore, Aquamandix (commercial filter medium consisting mainly of manganese and some silica) and virgin sand for adsorptive manganese removal were investigated and their manganese adsorption capacities and rates established. The effect of oxic and anoxic conditions and pH, were determined. The adsorption processes and mechanisms were investigated using three different kinetics models. The autocatalytic removal of manganese through adsorption at pH 6 and 8 was also investigated.

## 4.2    Theoretical Background

### 4.2.1 Oxidation of manganese

Manganese removal from groundwater is typically achieved through aeration followed by filtration. Basically in the filters, the dissolved Mn (II) is first adsorbed onto the filter media and then gets oxidized thereby creating new and more active adsorption sites that enhance further manganese removal. The oxidation of dissolved manganese in aqueous solution by dissolved oxygen follows the general equation (Stumm and Morgan 1996):

$$\frac{d[Mn\ (II)]}{dt} = - k_o[Mn\ (II)] + k_1[Mn\ (II)][MnO_x] \tag{4.1}$$

This equation shows that solid manganese oxide is required for the autocatalytic heterogeneous oxidation of manganese. In aqueous solutions homogenous manganese oxidation by oxygen has been reported to be very slow at pH values below 9 (Diem and Stumm, 1984; Stumm and Morgan 1996). In many conventional water treatment processes, chemical oxidants e.g. $Cl_2$, $ClO_2$, $KMnO_4$, $O_3$, NaOCl etc. are applied to enhance manganese oxidation over a wide range pH (i.e. pH 5 – 8) (Faust and Aly, 1998). The combination of green sand and potassium permanganate regeneration is commonly applied in U.S.A. The application of most oxidants might present some problems by giving rise to residuals and by-products like trihalomethanes that could pose a health hazard (Gallard and von Gunten, 2002).

Chapter 3 indicates that manganese adsorbed onto the surface of solid media e.g. IOCS may undergo oxidation in the presence of oxygen at lower pH values (i.e. pH < 9) without any use of chemicals. This mechanism of oxidation of adsorbed manganese by dissolved oxygen however needs to be investigated further to evaluate the mechanism that can promote the manganese oxidation at a pH lower than 9, since the rate of oxidation of Mn(II) in water at pH values below this level is extremely low. Moreover that study was limited to the use of only iron oxide coated sand for the investigation into the manganese adsorption phenomenon. Other filtering media with varying composition are expected to offer different and possibly a better adsorption potential than the iron oxide coated sand.

## 4.2.2 Isotherms of manganese adsorption on media

The affinity of the adsorbate for an adsorbent is quantified using adsorption isotherms, which are used to describe the amount of adsorbate that can be adsorbed onto an adsorbent at equilibrium and at a constant temperature. The Freundlich and Langmuir models are widely used mathematical descriptions of adsorption isotherms in aqueous systems. The empirical Freundlich model based on adsorption onto a heterogeneous surface is given as (see Chapter 3, section 3.2.3):

$$q_s = Kc_s^{1/n}$$

The Langmuir equation is based on the assumption that maximum adsorption corresponds to a saturated monolayer of adsorbate on the adsorbent surface, that energy of adsorption is constant and that there is no transmigration of adsorbate in the plane of the surface. The Langmuir equation is given as follows:

$$q = \frac{bq_m c_s}{1 + bc_s} \qquad (4.2)$$

Where $q_m$ is the maximum capacity of the sorption (mg/g) and b, the Langmuir adsorption constant is a constant related to the affinity of the binding sites (l/mg). $c_s$ is the equilibrium concerntration of adsorbate in solution (mg/l) and q is the amount of adsorbate adsorbed per unit mass of the adsorbent (g/g).

## 4.2.3 Kinetics of manganese adsorption

The kinetics of the manganese adsorption for the various filter media investigated were evaluated using the Lagergren equation derived from the Linear Driving Force model (LDF) and Potential Driving Second Order Kinetic model (PDSOK) (see Chapter 3). These two models are represented with the following rate equations below:

$$\ln [(q_s - q) / q_s] = - k_2 t \qquad (4.3)$$

$$\frac{t}{(c_o - c)} = \frac{1}{k_3(c_o - c_s)^2} + \frac{t}{(c_o - c_s)} \qquad (4.4)$$

where $q_s$ and q are the amounts of $Mn^{2+}$ adsorbed at equilibrium and at time t per unit mass of adsorbent. $c_o$ and c are the $Mn^{2+}$ concentrations in the liquid phase at time $t_0$ and time t respectively; $c_s$ is the $Mn^{2+}$ in the liquid phase at equilibrium; $k_2$ and $k_3$ are the pseudo first and second order rate constants of adsorption respectively.

The Dubinin-Kaganer-Radushkevisch model (DKR) (Erdem et al, 2004) has been applied to determine the sorption energy and subsequently to ascertain the prevailing adsorption mechanism. The DKR equation has the form:

$$\ln q = \ln X_m - \beta \varepsilon^2 \qquad (4.5)$$

where $X_m$ is the maximum sorption capacity and may represent the total specific micro pore volume of the adsorbent, $\beta$ is the activity coefficient related to mean sorption energy, and $\varepsilon$ is the Polanyi potential. The value of $\varepsilon$ is related to the equilibrium concentration ($c_S$) of the adsorbate at a temperature T (°K) by the relationship:

$$\varepsilon = RT \ln(1 + 1/c_S) \qquad (4.6)$$

where R, is the gas constant ($kJmol^{-1}K^{-1}$). With the value of $\beta$ determined from a plot of ln q versus $\varepsilon^2$, the sorption energy (E) can be worked out using the relationship:

$$E = 1/\sqrt{-2\beta} \qquad (4.7)$$

## 4.3    Experimental Section

### 4.3.1 Materials

For the equilibrium and kinetics experiments, eight various adsorbents including laterite, an iron-ore (Fe-ore), two different types of iron-oxide coated sand - one with high manganese (IOCS) and the other with low manganese content (LmIOCS), manganese green sand (original - MGS and regenerated forms - RMGS), Aquamandix (AQM, - a commercial filtering material) and virgin sand were investigated (see photographs of the various media included in this research in Appendix 4.1). The laterite and iron ore are virgin samples of natural, locally existing rocky deposits of Nyinahin and Shiene suburbs in the Ashanti and Northern regions respectively of Ghana. The laterite is basically hardened iron-rich mottled clay. The laterite possesses a continuous hard fabric of iron concretions wholly or partly joined together with rounded cavities and has aluminium, quartz and low percentage of silica (Raychaudhuri, 1980). The iron mineral present in the iron ore is mainly haematite ($Fe_2O_3$) and limonite ($2 Fe_2O_3.3H_2O$) with traces of pyrite (Minerals commission – Ghana, 1992).

The iron-oxide coated media were by-products obtained from the Noord-Bargeres (IOCS) and Brucht (LmIOCS) groundwater treatment plants in the Netherlands. The two iron-oxide coated sands had different mineral composition. The LmIOCS had a significantly lower content of manganese in its mineral coating. The manganese green sand is a commercial filter media formulated from glauconite. The glauconite component is a crystalline iron silicate mineral. In the course of the experiments it became necessary to regenerate the manganese greensand as is normally done in practice. For this purpose about 100 g of the manganese green sand was thoroughly washed with de-mineralized water to remove foreign material and subsequently immersed in 500 ml of 2 g $KMnO_4$ / l solution for 30 minutes. Afterwards the regenerated manganese green sand was washed with demineralized water and air-dried at 25°C for 3 days.

The Aquamandix consists predominantly of crushed manganese dioxide (78%) and generally lower percentages (i.e. < 6% by weight) of ferric oxide, silica and alumina (Aqua-techniek, 2007). The eight selected adsorbents were crushed and screened to reach particle size of < 75 µm.

## 4.3.2 Methodology

### 4.3.2.1 Batch experiments

Batch adsorption experiments with model water and selected adsorbents were done under oxic and anoxic conditions at pH 6 and 8. A stock solution of concentration 1000 mg Mn (II) /l was prepared using analytical-grade $MnSO_4.H_2O$ and acidified with concentrated HCl (32% assay) to a pH < 2. Experiments were performed with initial Mn (II) concentrations of up to 10 mg/l to prevent manganese carbonate precipitation (Chapter 3) and simulate the range of manganese levels in ground waters. Most ground waters have Mn (II) concentrations in the range of up to 2 mg/l however, on rare occasions Mn (II) concentrations could go up to about 10 mg/l. To ensure a stable pH during experiments, $NaHCO_3$ was added to model water. A relatively high dosage of $NaHCO_3$ was applied at pH 6 (Table 4.1) to achieve a higher buffering; this was not the case for pH 8 batch experiment to avoid possible manganese carbonate precipitation (Chapter 3).

**Table 4.1** Model water composition and applied conditions for batch adsorption experiments.

|  | Oxic experiment | | Anoxic experiment | |
|---|---|---|---|---|
|  | pH 6 | pH 8 | pH 6 | pH 8 |
| [$NaHCO_3$] (mg/l) | 1000 | 84 | 1000 | 84 |
| Initial [$Mn^{2+}$] (mg/l) | 2 | 0.8 | 0.1 - 8 | 0.1 – 0.8 |

For the oxic experiments acid-cleansed 500 ml plastic bottles fitted with tubes for periodic sampling were filled with de-mineralized water dosed with the appropriate amount of $NaHCO_3$ (Table 4.1) and their pH subsequently adjusted with 6N HCl or 1N NaOH solutions to the required level. After thorough mixing of the prepared model water and spiking targeted Mn (II) concentration, 4 g of pulverized filter media were added. Bottles were kept at 20 ± 1 °C and placed on a shaker. Blank tests were carried out without the addition of adsorbent.

Adsorption isotherms under anoxic conditions were established to classify the
adsorbents as a function of their manganese adsorption capacity at equilibrium.
Experiments under anoxic conditions were done in 1.5 L glass reactors with model water
(Table 4.1) and adsorbent dosages ranging from 0.25 to 8 g pulverized filter media /l
(see Chapter 3). The reactor had ports to allow for sampling, dissolved oxygen
determination, temperature and pH measurement, gas supply and mechanical stirring.
Nitrogen gas was infused into the reactor when required to attain and maintain anaerobic
conditions. Carbon dioxide was introduced for pH adjustment. These gas infusions and
continuous stirring ensured proper mixing. Periodic sampling was done at regular time
intervals to determine the rate of adsorption and equilibrium conditions. Equilibrium was
considered to have been reached when the difference in manganese concentration of two
consecutive samples taken over a period of 10 hrs was $\leq$ 0.02 mg/l. All experiments
were run in duplicate and samples filtered through a 0.45 $\mu$m membrane filter using a
polypropylene syringe filter. Manganese analysis was done using Perkin Elmer 3110
spectrometer in accordance with the Dutch Standard Method NEN 6457. The method
used has a detection level of 0.02 mg Mn (II)/ l.

### 4.3.2.2 Models applied
The Langmuir and Freundlich's isotherms together with the Linear Driving Force
(Lagergren equation), the Potential Driving Second Order Kinetic and the DKR models
were applied to the data obtained from the batch experiments to provide information on
the mechanism of adsorption and the rate of adsorption.

### 4.3.2.3 Characterization of selected media - chemical composition
To determine the iron and manganese content of the various filter media, a mixture of
each filter media comprising 1g of the pulverized filter media, 10 ml of analytical grade
concentrated $HNO_3$ (65%) and topped up to 50 ml with demineralized water was
allowed to stay overnight. Subsequently another 5 ml aliquot of the $HNO_3$ was added
and the adsorbent-acid mixture heated on a hot plate at a temperature of 200 – 300 °C
for about 2 hours avoiding splattering. The heating was terminated when a clear solution
has been obtained from the heat digestion. The digested sample was filtered and the
filtrate diluted with demineralized water in a 250 ml volumetric flask. The resulting
solution was analyzed with atomic absorption spectrometer for the iron and manganese
content of the filter media (Standard methods, 2005).

# 4.4    Results and Discussion

## 4.4.1 Adsorption isotherms
The anoxic manganese adsorption data for the filter media tested fitted well the
Freundlich isotherm at pH 6 and 8 (Figure 4.1). As expected, the manganese adsorption
at pH 8 was higher than at pH 6 for all the media. The manganese green sand (MGS) and
virgin sand did not show manganese adsorption at pH 6. The MGS rather leached
manganese into the aqueous solution. To investigate and confirm this manganese
leaching, the manganese green sand was regenerated with 2 g $KMnO_4$/l and (re)used for

the manganese adsorption test under identical conditions. The regenerated manganese green sand (RMGS) also leached manganese and did not show any manganese adsorption at pH 6. Aquamandix exhibited the highest manganese adsorption capacity at both pH levels investigated. The adsorption capacities at pH 8 decreased in the order: Aquamandix > IOCS > MGS > LmIOCS > Laterite > Iron-ore > Virgin sand at pH 8 as in Figure 4.2.

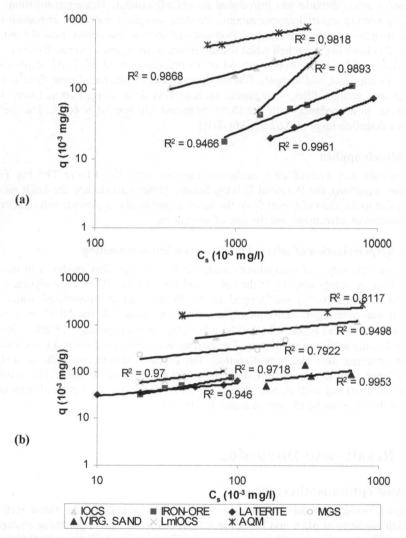

(a)

(b)

**Figure 4.1** Freundlich's manganese adsorption isotherms determined under anoxic conditions for the various pulverized media at pH 6 (a) and pH 8 (b). Model water: 1000 mgNaHCO$_3$ /l for pH 6 and 84 mg NaHCO$_3$ for pH 8.

Chapter 4: Manganese (II) adsorption characteristics of selected filter media for groundwater
treatment: equilibrium and kinetics

85

The two iron oxide coated sand media (i.e. IOCS and LmIOCS) gave different
manganese adsorption patterns. These two media had comparable iron content but
different manganese content in their composition (i.e. IOCS → 24.5 mg Mn/g; LmIOCS
→ 4.66 mg Mn/g) (Table 4.4). From Figures 4.1 & 4.2, it can be seen that IOCS
exhibited considerably higher manganese adsorption at pH 6 and 8 than LmIOCS.

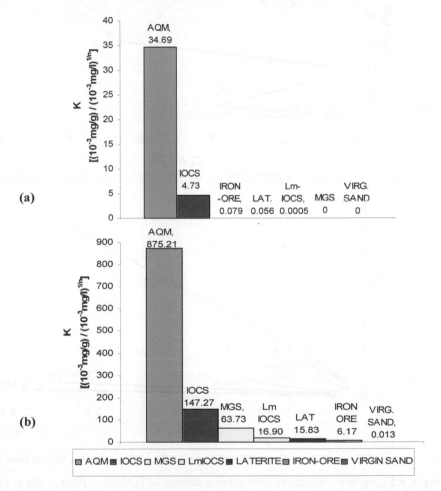

**Figure 4.2** Effect of pH on manganese adsorption capacities of selected pulverized
media under anoxic conditions at pH 6 (a) and pH 8 (b).

LmIOCS showed very low adsorption capacity especially at pH 6, indicating that
manganese adsorption is enhanced by the presence of manganese in the mineral coating
of the media. The manganese adsorption capacities of the Aquamandix and IOCS
increased by 25 – 30 times with an increase in pH from 6 to 8 (Figure 4.2). The
experimental data for most of the filter media fitted also the Langmuir isotherm

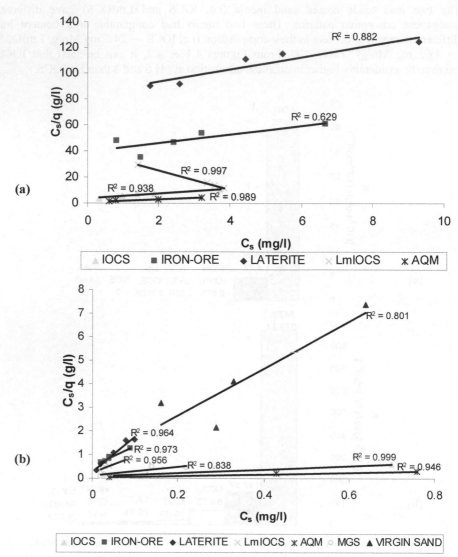

**Figure 4.3** Langmuir manganese adsorption isotherms of pulverized filter media at pH 6 (a) and pH 8 (b) under anoxic conditions. Model water: 1000 mg $NaHCO_3$ /l for pH 6 (a) and 84 mg $NaHCO_3$ /l for pH 8 (b).

($R^2 > 0.93$) at pH 8, except for MGS and virgin sand (Figure 4.3 and Table 4.2). At pH 6 only data obtained for Aquamandix and IOCS fitted well the Langmuir isotherm under anoxic conditions and demonstrated increasing manganese adsorption when the concentration increases. The Langmuir parameter $q_m$ (i.e. maximum capacity) is also a measure for comparing the adsorption potential of different adsorbents for the same adsorbate. Table 4.2 and Figure 4.3 confirm that the Aquamandix and IOCS have higher

Chapter 4: Manganese (II) adsorption characteristics of selected filter media for groundwater
treatment: equilibrium and kinetics

87

adsorption capacities than all the adsorbents tested under anoxic conditions at pH 6 or 8.

**Table 4.2** Parameters of the Langmuir manganese isotherms for the various filter media.

| Filter media | pH 6 | | pH 8 | |
|---|---|---|---|---|
| | $R^2$ | $q_m$ | $R^2$ | $q_m$ |
| AQM | 0.9891 | 0.99 | 0.9461 | 2.69 |
| IOCS | 0.9375 | 0.51 | 0.9985 | 1.36 |
| MGS | - | - | 0.8383 | 0.52 |
| LmIOCS | 0.9974 | -0.13 | 0.9558 | 0.15 |
| Fe-ore | 0.629 | 0.30 | 0.9727 | 0.12 |
| Virgin sand | - | - | 0.8008 | 0.10 |
| Laterite | 0.8824 | 0.21 | 0.9643 | 0.07 |

## 4.4.2 Manganese removal under oxic and anoxic conditions

Batch adsorption experiments performed under oxic conditions at pH 6 and 8 showed a trend of rapid drop in the residual manganese (II) concentration in solution during the period of the first hour followed by a slow manganese removal phase (Figure 4.4). The initial rapid manganese adsorption could be due to availability of easily accessible surface area of the adsorbents. The following slow manganese removal phase may be due to slow in-pore diffusion or continuous adsorption as a result of oxidation of the adsorbed manganese. As expected manganese removal was more pronounced at the higher pH of 8 than pH 6 for all the media tested (Figure 4.4).

Both the original manganese greensand (MGS) and the regenerated manganese green sand (RMGS) did not show any manganese removal but rather leached manganese into the aqueous solution at pH 6 (Figure 4.4). However, these manganese green sands at pH 8 demonstrated appreciable adsorption potential. The order of the extent of manganese removal for the various filter media after the seven days experimentation is as follows:

At pH6: AQM > IOCS > LmIOCS > Iron-ore > Laterite (LAT) > Virgin sand;
At pH8: AQM > IOCS > LmIOCS > RMGS > Iron-ore > MGS > LAT > Virgin sand.

The manganese reduction from the solution at pH 6 with the IOCS and Aquamandix proceeded perpetually under oxic conditions, suggesting that oxidation of the adsorbed manganese may be occurring. From literature oxidation of manganese (II) is not expected to occur at the pH values below 9 in solutions, however here in the presence of the adsorbent oxidation is most likely to occur. If the oxidation of adsorbed manganese ion is occurring, then the oxidized adsorbed manganese is likely to become a new site for attachment of dissolved $Mn^{2+}$ ions. In this way the sites (partially) occupied by $Mn^{2+}$ are regenerated. It is expected that the rate of oxidation / regeneration will be higher at pH 8 than at pH 6.

To investigate and confirm the possibility of oxidation of adsorbed $Mn^{2+}$, batch reactors experiments with Aquamandix grains were set up using a similar modeled water composition under both anoxic and oxic conditions at pH 8. Under oxic conditions, the removal rate was higher and 94 % manganese removal compared to 74% for the anoxic condition was noted after 7 days (Figure 4.5). The recorded higher adsorption potential and continuing adsorption of Aquamandix under the oxic conditions is attributed to the adsorption of manganese and subsequent oxidation, creating new adsorption sites.

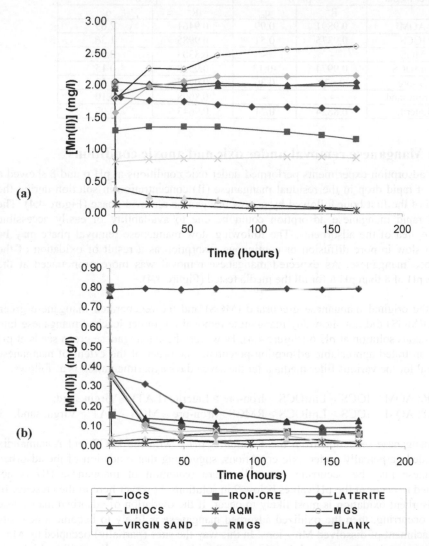

**Figure 4.4** Manganese adsorption onto pulverized filter media under oxic conditions at pH 6 (a) and pH 8 (b). Model water: 1000 mgNaHCO₃/l, for pH6 and 42 mgNaHCO₃/l for pH 8; initial Mn (II) concentration 2 mg/l at pH 6 and 0.8 mg/l at pH 8; Dosage of pulverized media: 8g/l for pH 6 and 4 g/l for pH 8.

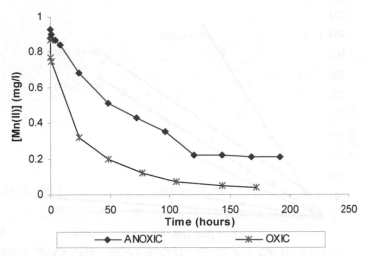

**Figure 4.5** Effect of dissolved oxygen on manganese removal by Aquamandix grains
(0.4 g/l) at pH 8.

## 4.4.3 Adsorption kinetics

The Linear Driving Force model (Lagergren equation) did not provide a very good fit
over the whole range of the contact time for all the filter media tested (apart from AQM,
all the media tested had $R^2 \ll 0.9$) (Figure 4.6).

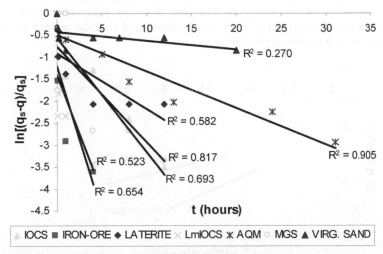

**Figure 4.6** Kinetics model: Linear Driving Force (Lagergren equation) for manganese
adsorption data obtained under anoxic conditions at pH 8.

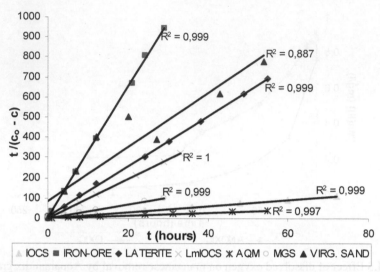

**Figure 4.7** Kinetics model: Potential Driving Second Order Kinetic model for manganese adsorption data obtained under anoxic conditions at pH 8.

The Potential Driving Second Order Kinetic (PDSOK) rate model adequately describes the kinetics of the manganese adsorption with a good correlation ($R^2 > 0.99$) for all the filter media tested except the virgin sand (Figure 4.7). According to literature (Ho et al, 1995; Wankasi et al, 2006), if a linear relationship is obtained for the plot of $t/(c_o-c)$ against t (as in Figure 4.7) the sorption process may be described as chemi-sorption. Therefore with $R^2 > 0.99$ obtained from this study it can be suggested that the Pseudo Second Order adsorption mechanism is predominant and that the overall manganese adsorption is likely a chemi-sorption process.

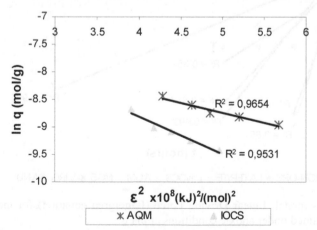

**Figure 4.8** DRK model applied with manganese adsorption data obtained under anoxic conditions.

Chapter 4: Manganese (II) adsorption characteristics of selected filter media for groundwater treatment: equilibrium and kinetics

91

The Figure 4.8 and Table 4.3 depict the Aquamandix and the IOCS having high sorption energy values. According to literature (Rieman and Walton 1970; Erdem et al. 2004), sorption energies of 8 – 16 kJ/mol are the orders for an ion-exchange mechanism. Therefore the range of sorption energies obtained here (i.e. 9000 – 12000 kJ/mol) may suggest that the mechanism prevailing may not be ion-exchange mechanism. The DKR model fits the experimental data reasonably, however not as good as the Potential Driving Second Order model.

**Table 4.3** Parameters obtained from computation and DKR plot (Figure 4.8).

|  | AQM | IOCS |
|---|---|---|
| $X_m$ (mmol/g) | 0.966 | 2.223 |
| $\beta$ (mol$^2$/kJ$^2$) | $-3.60 \times 10^{-9}$ | $-6.76 \times 10^{-9}$ |
| Sorption energy (E, kJ/mol) | 11785 | 8600 |
| $R^2$ | 0.9654 | 0.9531 |

### 4.4.4 Relationship between the manganese adsorption potential and the media characteristics

From Table 4.4 it is observed that, for the media AQM, IOCS, MGS and LmIOCS a high adsorption potential coincides with the manganese content. The other media having much lower manganese content do not show this phenomenon. LmIOCS has much lower manganese content than IOCS, suggesting that manganese presence in the coating is essential for a high manganese adsorption capacity.

**Table 4.4** The chemical composition and characteristics of the various media.

| Media | Fe content (mg/g media) | Mn content (mg/g media) | Mn/Fe | $q_m$ | K | $k_2$** |
|---|---|---|---|---|---|---|
| **Aquamandix** | 43.00*** | 780*** | 16.67 | 2.69 | 875.20 | 0.08 |
| **IOCS** | 509.00 | 24.50 | 0.05 | 1.36 | 147.27 | 0.22 |
| **MGS** | 42.49 | 7.50 | 0.18 | 0.52 | 63.73 | 0.26 |
| **LmIOCS** | 401.87 | 4.66 | 0.01 | 0.15 | 16.90 | 0.54 |
| **IRON-ORE** | 0.68 | 0.05 | 0.07 | 0.12 | 6.17 | 0.67 |
| **Laterite** | 3.03 | 0.08 | 0.03 | 0.07 | 15.83 | 0.13 |
| **Virgin Sand** | 6.21 | 0.11 | 0.02 | 0.10 | 0.01 | 0.02 |

$q_m$ values quoted from table 4.2; **slope values derived from figure 4.6; *** Values adapted from Aqua-Techniek (2001); K is the adsorption capacities from Figure 4.2.

The levels of k-values derived from the PDSOK graph coincide with the adsorption potential. This observation suggests that the internal mass transfer is related to the adsorption capacity.

## 4.5  Conclusion

- The Aquamandix, IOCS, LmIOCS, Iron-ore and Laterite demonstrated the potential to adsorb manganese at pH 6 and 8 under both oxic and anoxic conditions. Manganese green sand showed manganese adsorption capacity only at pH 8. The manganese adsorption capacities of all the media tested were much higher at pH 8 than pH 6.

- The manganese adsorption capacities of the various media were found to be in the following order:

At pH 6:
AQM > IOCS > LmIOCS > Fe-ore > LAT > Virgin sand;
At pH 8:
AQM > IOCS > LmIOCS > RMGS > Fe-ore > MGS > LAT > Virgin sand.

Aquamandix demonstrated the highest adsorption capacity and much higher than sand. Consequently Aquamandix qualifies as a potential alternative for the sand that is commonly used in manganese removal filters in practice.

- Iron oxide coated sand, containing sufficient manganese is the second best option with a reasonably high adsorption capacity and being a by-product of the water treatment plants could be the cheapest substitute. Virgin sand has by far the lowest adsorption capacity.

- The adsorption potential of Aquamandix, iron oxide coated sand (with high and low manganese content) and manganese green sand, coincides with manganese content, indicating that manganese is essential for adsorption capacity of media.

- The adsorption kinetics of manganese under anoxic condition on different media gave the best fit with the Potential Driving Second Order Kinetic (PDSOK) model and the DKR model.

- The linear relationship obtained for the PDSOK model and the high sorption energy obtained from the DKR model indicate that the manganese adsorption onto the Aquamandix and IOCS follows the chemi-sorption mechanism.

- Aquamandix media demonstrated autocatalytic adsorption under oxic conditions and pH 8. Results obtained with other media were less pronounced and need to be studied more in detail.

## 4.6    References

Aqua-Techniek (2007) Aquamandix:
        http://www.aqua-techniek.com/html/filtermedia.htm. (Last accessed - September, 2008).
Barua, R., (2006) 'Long term monitoring and optimization of manganese removal with IHE family
        filter' M.Sc. Thesis, UNESCO - IHE, Institute for Water Education, The Nederlands.
Diem, D. and Stumm, W. (1984) Is dissolved $Mn^{2+}$ being oxidized by $O_2$ in the absence of Mn-
        Bacteria or surface catalysts? *Geochim Cosmochim Acta* 1984; 48:1571 – 3.

Erdem, E., Karapinar, N., Donat, R. (2004) The removal of heavy metal cations by natural
zeolites. *J. of Colloid and Interface Science* 280; 309 – 314.

Faust, S. D. & Aly, O.M. 1998 Chemistry of water treatment: 2$^{nd}$ Edition. Ann Arbor Science
book. Lewis Publishers – CRC Press LLC, Boca Raton, London, New York,
Washington D.C. pp 145 – 146.

Gallard, H.U. and von Gunten U. (2002), Chlorination of natural organic matter: kinetics of
chlorination and of THM formation. *Water Res.* 36:65 – 74.

Ho, Y.S.; Wase, D.A.J. and Forster, C.F. (1995) Batch nickel removal from aqueous solution by
sphagnum moss peat. Water Research, 29:(5): 1327 – 1332.

Katsoyiannis, I.A. and Zouboulis, A.I. (2003) Biological treatment of Mn(II) and Fe(II) containing
groundwater: kinetic considerations and product characterization. *Water Research*
38:1922 – 1932.

Minerals Commission Publication Project – Ghana (1992) Report on the Shiene Iron Ore Deposits.
Ghana Geological Survey Department; Archive Report No. 85.

Oscarson, D.W., Huang, P.M., Defosse, C.,and Herbillon, A. (1981) Oxidative power of Mn(IV)
and Fe(III) oxides with respect to As(III) in terrestrial and aquatic environments.'
*Nature*, 291: 50 – 51.

Po-Yu Hu, Yung-Hsu, H. Jen-Ching, C. and Chen-Yu, C. (2004) Adsorption of divalent
manganese ion on manganese-coated sand. *Journal of Water Supply: Research and
Technology - AQUA* 53.3: 151 – 158.

Raychaudhuri, S.P. (1980) The occurrence, distribution, classification and management of laterite
and lateritic soils. *Cah.O.R.S.T.O.M., ser. Pedol.* vol. XVIII, No. 3-4: 249-252.

Rieman, W. and Walton, H. (1970) Ion exchange in Analytical Chemistry, in: International Series
of Monographs in Analytical Chemistry, Vol. 38, Pergamon, Oxford.

Scott, M.J. and Morgan, J.J. (1995) 'Reaction at oxide surfaces. I:Oxidation of As(III) by synthetic
birnessite.' *Environmental Sci. Technol.* 29(8), 1898 – 1905.

Stumm, W. and Morgan, J.J. 1996 Aquatic Chemistry: Chemical equilbria and rates in natural
waters, 3$^{rd}$ Edition. Wiley, New York. Pp 163 – 172, 684 – 686.

Stembal, T., Markic, M., Briski, F. and Sipos, L. (2004) Rapid start-up of biofilters for removal of
ammonium, iron and manganese from groundwater. *Journal of Water Supply; Research
and Technology – AQUA* 53.7: 510 – 518.

Standard methods for the examination of water and wastewater (2005). American Public Health
Association / American Water Works Association / Water Environment Federation,
Washington, DC.

Takamatsu, T..k Kawashima, M., and Koyama, M. (1985). The role of Mn$^{2+}$ - rich hydrous
manganese oxide in the accumulation of arsenic in lake sediments. *Water Res.,* 19(8),
1029 – 1032.

Wankasi, D., Horsfall Jnr, M. and Ayabaemi, I. S. (2006) Sorption kinetics of Pb2+ and Cu2+ ions
from aqueous solution by Nipah palm (Nypa fruticans Wurmb) shoot biomass.
Electronic Journal of Biotechnology – ISSN: 0717-3458; vol. 9 No. 5. 587 – 592.

WHO, (2004) Guidelines for drinking – water quality. Third edition. Vol. 1 recommendations 397
– 399 & 492.

Yan-Chu (1994) 'Arsenic distribution in soils.' Arsenic in the environment. Part 1: Cycling and
characterization, J.O.Nriagu ed., Wiley, New York, 17 – 49.

# APPENDIX 4.1

Photograph of the six different media studied

(a) Aquamandix (particle size 0.5 – 1.0 mm);
(b) IOCS (particle size: 2.0 – 3.15 mm);
(c) Manganese greensand (particle size (0.5 -1.5 mm);
(d) A stone of iron ore which was crushed and pulverised to particle size < 75μm;
(e) A Laterite stone which was crushed and pulverised to particle size <75 μm;
(f) Virgin sand (particle size 0.5 – 1.2 mm).

# CHAPTER FIVE

# MANGANESE REMOVAL FROM GROUNDWATER; PROBLEMS IN PRACTICE AND POTENTIAL SOLUTIONS

This Chapter has been published as:
Buamah, R. Petrusevski, B. de Ridder, D. van de Wetering, T.S.C.M. and Schippers, J.C. (2009) Manganese removal in groundwater treatment: practice, problems and probable solutions. *Journal of Water Science and Technology* 9.1: 89-98.

and presented as:

Buamah, R. Petrusevski, B. and Schippers, J.C. (2008) Manganese removal in groundwater treatment: practice, problems and probable solutions. In: Proceedings of *the IWA - 5th Leading Edge Technologies Conference at Zurich* 1 – 4 June, 2008.

## Abstract

Many drinking water production plants use rapid sand filters for the removal of manganese from groundwater. The ripening and start-up of manganese removal on newly installed sand media is slow, taking several weeks till months. Reducing this period in order to prevent the loss of water during this phase has become an issue of concern. It has been reported that sand filter media in some manganese removal filters has to be replaced after only a few years of operation due to fast manganese breakthrough.

In this study, pilot and bench scale experiments were conducted to investigate the mechanism, influence of operational conditions and measures that enhance manganese removal capacity of the sand media.

The development of the adsorptive/catalytic coating on the sand media in the filters of a pilot plant was very slow, notwithstanding the relatively high pH of 8. Increasing the level of manganese in the feed water during the ripening period accelerated the start up of the manganese removal. Higher rate of filtration resulted in a longer start up period, most likely due to more frequent backwashing. More frequent backwashing is expected to remove a (larger) part of the newly formed manganese oxide catalyst.

Bench scale tests under oxic conditions carried out on the sand samples taken from the top of the filters, after 15 weeks of operation, showed continuous adsorption / oxidation of manganese. However, the manganese removal achieved in the batch adsorption experiments was too low to explain the complete removal of manganese in the pilot filters. Covering of the catalyst by iron, specifically at the top of the filter media and / or aging of the catalyst between sampling and measuring of the capacity might be the reasons. This developed bench scale test is a promising tool in determining the manganese removal of filter media.

Scanning Electron Microscopic (SEM) investigations conducted on the surface of the sand grains taken after 15 weeks of pilot filter operation, indicated that part of the grains surface consist substantially of manganese. The spectrum obtained from the examination of different micro locations on the grains surface showed irregular coating and development of the manganese oxide catalyst.

Determined Freundlich adsorption isotherms of different media, measured under anoxic conditions, demonstrated that a manganese bearing mineral (Aquamandix) has by far the highest adsorption capacity while virgin quartz sand has the lowest capacity. Consequently Aquamandix is a promising alternative filter medium in solving slow start up problems in manganese removing filters.

## 5.1    Introduction

Groundwaters generally contain one or more contaminants like iron, manganese, ammonium, methane and natural organic matter e.g. humic acid, etc. For drinking water supply purposes, these contaminants need to be removed or reduced to acceptable levels. High manganese levels in drinking water can give rise to aesthetical and health problems. Together with dissolved iron, manganese (II) could form deposits of iron hydroxide and manganese oxides in distribution lines. If the flow in the distribution lines increases (e.g. during daily peak demand), the sediment can become re-suspended and may result in customer complaints due to incidents of 'black' or 'brown' water. These 'black water' incidents may occur when manganese concentrations in treated water is as low as 0.02 mg/l (MWH, 2005). To exclude the possible health hazard as a result of chronic exposure to manganese, a health-based guideline of 0.4 mg/l for drinking water has been recommended (WHO, 2006).

In ground water treatment two processes are frequently applied namely:

- aeration followed by rapid sand filtration;
- filtration through Green Sand (with or without preceded aeration) with regeneration (oxidation of adsorbed iron and manganese) by an oxidant e.g. potassium permanganate.

Many groundwater treatment plants (GWTP), in Europe and other parts of the world, aerate the raw water to remove methane and introduce oxygen to oxidize Fe (II) and Mn (II) to their respective oxides; in addition adsorbed iron and manganese may be oxidized as well. Ammonium can be removed biologically in the following sand filtration. The oxidized iron is removed by physical filtration and the dissolved Fe (II) and Mn (II) are adsorbed on the filter media and subsequently oxidized. Humic acids may combine with Fe (II) and Mn (II) to reduce the rate of manganese oxidation. The extent and rate of oxidation in a filter is dependent on the pH, the dissolved oxygen content, presence of iron and manganese oxide coatings on the filter media, the GWTP design and the operational conditions (i.e. the backwashing regime employed).

Frequently, it has been observed that, the manganese removal capacity of the sand filters reduces after a couple years in operation. As part of the measures to solve this problem, the filter media has to be regularly replaced with new sand. It is not yet clear which mechanisms are responsible for this phenomenon. Covering of the manganese oxide coating by ferric hydroxides and / or bacteria might be one of the reasons. After replacements, a slow start-up of manganese removal has typically been observed. This study is focused on the manganese removal problems from ground water, with the aim to get insight in the mechanisms governing the start up of manganese removal in rapid sand filters and to indicate potential solutions. For this purpose a pilot plant study has been conducted at the GWTP Haaren of Water Supply Company Brabant. Pilot rapid sand filters were operated at different filtration rates and manganese concentrations in the influent.

Laboratory and bench scale investigations were done to determine manganese adsorption capacities of the media after several months of operation. These tests were done under oxic and anoxic conditions to differentiate between adsorption capacity and adsorption / catalytic oxidation capacity. In addition Scanning Electron Microscopy investigations on the grains of the filter media were conducted to determine the trend and style of the coating development.

## 5.2     Theoretical background

During the operation of rapid sand filters, the influx of Fe (II) and Mn (II) ions from the raw water facilitates formation of iron oxide coating on the sand grains. The coatings formed consist predominantly of iron-(hydro)-oxides and normally have manganese oxides embedded within. These iron and manganese oxides (e.g. $Fe_2O_3$, $FeOOH$, $Mn_3O_4$, $MnO_2$) in the sand coating enhance adsorption of in-coming Fe (II) and Mn (II) ions. Certain oxides in the coating especially freshly formed manganese oxide like $MnO_2$ and $Mn_3O_4$ catalyse the oxidation of adsorbed Mn (II) in the presence of dissolved oxygen and favourable pH (Stembal et al. 2004). Consequently new adsorption sites are created thus preventing saturation of adsorption capacity. The presence of high amount of these catalytic oxides on the sand grains is therefore paramount for the rapid start-up of manganese removal in the sand filters.

Literature informs that the iron content of the raw water, pH and length of time in use contribute to the rate of growth of the iron oxide coating (Sharma, 1999). Despite these findings, the length of time required to achieving adequate oxidation of the adsorbed iron and manganese in particular manganese, to subsequently create new adsorption sites on the filter media still remains to be determined.

Manganese can exist in eleven oxidative states; the most environmentally and biologically important manganese compounds are those that contain Mn (II), Mn (IV), or Mn (VII) (US EPA, 1994). The equilibria between the various forms of Mn are determined mainly by pH and redox potential as illustrated in Figure 5.1. At low and neutral pH and low redox potential Mn (II) prevails whereas at high pH and low redox potential, the formation of $MnCO_3$ is favoured. Manganese cannot be oxidised and removed from water as readily as iron, so in some cases aeration-filtration alone is not adequate. As a consequence several groundwater treatment plants make use of an additional aeration-filtration stage. In the first stage primarily iron is removed and in the second stage mainly manganese is removed due to higher pH and absence of iron covering the manganese adsorbent/catalyst.

The manganese oxidation kinetic equation proposed by Stumm and Morgan (1996) is:

$$\frac{d[Mn(II)]}{dt} = - k_o [Mn(II)] + k_1[Mn(II)][MnO_2] \qquad (5.1)$$

where:

$$k_o = kPO_2 . [OH^-]^2 \qquad (5.2)$$

$k_o$ = reaction rate constant ($l^2/mol^2$.atm.min)
$k_1$ = reaction rate constant ($l^3/mol^3$.atm.min)
$PO_2$ = Partial pressure of oxygen (atm)

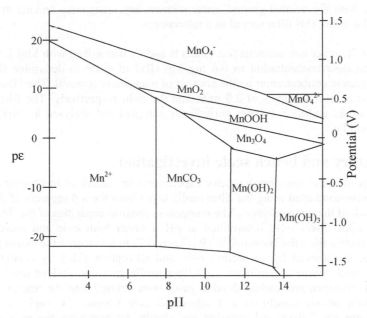

**Figure 5.1** pε – pH diagram for aqueous Mn (Stumm & Morgan, 1996).

So aside the pH, the Mn (II) and oxygen concentrations, the rate depends on the product of the [Mn (II)] and the [$MnO_2$], implying that the reaction is autocatalytic. At pH values below 9 the rate of oxidation in solutions is very low. In rapid (sand) filters manganese removal has been observed down to pH 6.9. This phenomenon is attributed to the catalytic effect of manganese oxides e.g. $Mn_3O_4$ and $MnO_2$. The mechanisms involved are expected to be as follows:

- adsorption of Mn (II) on the surface of the media e.g. sand;
- oxidation of adsorbed Mn (II) to $Mn_3O_4$ and $MnO_2$ that takes place at much lower pH values than 9 (i.e. down to pH 6.9);
- adsorbed and subsequently oxidized Mn (II) act as newly created adsorbent. The old and new adsorbents catalyse the oxidation of the subsequently adsorbed Mn (II).

## 5.3    Materials and methods

### 5.3.1 Pilot plant experiments

Pilot plant experiments were conducted using three filter columns (Internal diameter of 18 cm) and virgin sand (particle size 0.71 – 1.25 mm; bed depth of 75 cm) as filter media at the Haaren GWTP in the Netherlands. The three columns were run with the aerated ground water containing about 0.04 mg Mn (II)/l, 0.55 mg $NH_4^+$/l, 0.4 mg Fe (II)/l, 9.6 mg $O_2$/l, alkalinity of 183 mg$HCO_3^-$/l, and pH 8 (Figure 5.2). One column (i.e.

filter A) was fed with the aerated ground water without any manganese spiking at a filtration rate of 2.8 m/hr. This filter served as a reference.

The feed water of the other two columns (i.e. filters B and C) was spiked with $MnCl_2$ to increase the manganese concentration to 0.6 mg Mn (II)/l in order to determine the influence of manganese concentration on the start-up of manganese removal. The Filters B and C were run at filtration rates of 2.8 m/hr and 0.6 m/hr respectively. The filters were run for 22 weeks and their filtrates periodically sampled and analysed for $NH_4^+$, $NO_2^-$, $NO_3^-$, Mn and Fe.

## 5.3.2 Laboratory and bench scale investigations

After 15 weeks of the filter runs, batch reactor experiments (as described in Chapter 3, section 3.3.1.1) were conducted using the filter media taken from the top segment of the filter bed from each of the three columns. The manganese removal capacities of the filter media from the pilot filters were determined at pH 8 under both oxic and anoxic conditions using filtrate from the Haaren GWTP (Figure 5.2) to investigate the extent to which manganese is removed by adsorption only and adsorption aided by catalytic oxidation. In the batch reactor experiments 1 g/l filter media grains was added and 0.3 mg Mn/L spiked. Nitrogen and carbon dioxide gases were infused into the reactor to attain and maintain anoxic conditions and adjust pH (see Chapter 3). Each batch experiment was run for 7 days and sampled periodically to determine the residual dissolved manganese content. The samples were filtered through a 0.45 μm membrane filter and analysed for manganese using Perkin Elmer 3110 spectrometer having a detection level of 0.02 mg Mn (II)/l. The analysis was done in accordance with the

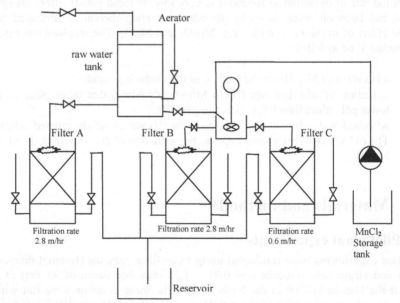

**Figure 5.2** A schematic diagram of the Haaren pilot set-up.

Dutch Standard Method NEN 6457. Equilibrium was considered to have been reached when the difference in manganese concentration of two consecutive samples taken over a period of 10 hrs was ≤ 0.02 mg/l. All experiments were run in duplicate.

### 5.3.3 Development coating and Scanning Electron Microscopy investigations

The filter media samples taken after the 15 weeks of pilot filter operation were investigated using Scanning Electron Microscopy (SEM) with a microscope JEOL JSM-6500F combined with an EDX probe. Electron beam energy of 15 – 20 keV and a beam current of 40nA were applied. The investigation was performed to identify the elemental composition at selected specific locations on the surface of the sample grains.

## 5.4 Results and discussions

### 5.4.1 Pilot plant experiments

All three filters demonstrated removal of Mn (II) after a certain ripening period. In general it took about 40 days before manganese removal started in the filters. Filter A (that was run at 2.8 m/hr; with 0.4 mgFe(II)/l, 0.04 mgMn (II)/l and no extra manganese dosage) took 100 days to achieve complete manganese removal. Filters B and C (running at 2.8 and 0.6 m/hr and both with an elevated manganese level of 0.4 mgFe(II)/l and about 0.6 mgMn(II)/l) arrived at that stage already after 79 and 51 days of operation respectively (Figures 5.3, 5.4 and 5.5).

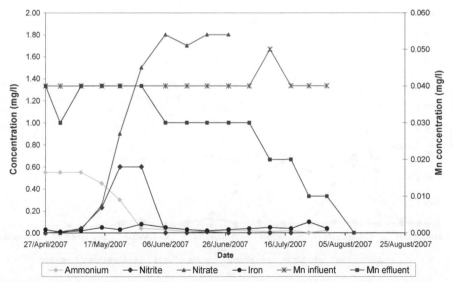

**Figure 5.3** Developments in the water quality of the effluent of filter A running at 2.8 m/hr with 0.04 mg Mn (II)/l and 0.4 mg Fe (II)/l.

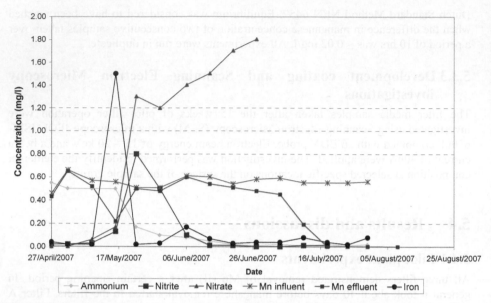

**Figure 5.4** Developments in the water quality of the effluent of filter B running at 2.8 m/hr with 0.6 mg Mn (II)/l and 0.4 mg Fe (II)/l.

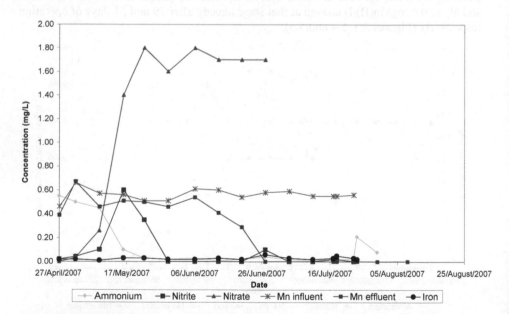

**Figure 5.5** Developments in the water quality of the effluent of filter C running at 0.6 m/hr with 0.6 mg Mn (II)/l and 0.4 mg Fe (II)/l.

The potential reasons for these pronounced differences are as follows:

- The built up of adsorptive catalytic manganese oxides is governed mainly by the concentration in the feed water and / or the total amount of manganese entering a filter. This effect may explain why filter B demonstrated a faster achievement of complete manganese removal than filter A (having low Mn(II) but same filtration rate). The manganese concentration in the feed water was 15 times higher in filter B. Assuming the manganese influx to be a governing factor of this, it would be expected that filter B would perform better than filter C. However, the results showed filter C doing better.
- The unexpected difference in behaviour of filter B and C might have been caused by a difference in the backwash regime e.g. due to the higher filtration rate, filter B has been backwashed more frequently than filter C. As a result more newly formed adsorptive catalyst might have been removed during backwashing to a larger extent than in filter C.
- Difference in bacterial growth due to ammonium removal might be another potential reason for the observed differences. However there is no pronounced difference in the trends of the start-up and complete ammonium removal. The formation and disappearance of nitrite might have an effect on the development of Mn (II) removal. The Figures 5.3, 5.4 and 5.5, demonstrate that Mn (II) removal starts after the disappearance of nitrite. Further research is however, necessary to verify whether nitrite plays a role in inhibiting / retarding the development of the manganese catalyst.

## 5.4.2 Laboratory and bench scale investigations

The bench scale batch experiments showed that, for the various filter media (i.e. filters A, B and C) the adsorption capacities of the grains do not differ much within the first 100 hours under both oxic and anoxic conditions (Figures 5.6, 5.7 and 5.8). Under the anoxic conditions, after the initial drop, the dissolved manganese concentration was found to have increased before leveling off (media of filter B, Figures 5.6b and 5.8a); this manganese release may probably be the dissolution of manganous carbonate that might have precipitated locally and gotten embedded within the mineral coating formed during the pilot experimentation. This might have been induced by a higher pH due to unintended additional aeration in the system. Using solubility product equations and Phreeqc software program, for pH 8 and a hydrogen carbonate concentration of 3 mmol/l, manganous carbonate precipitation commences when dissolved manganese concentration is > 0.3 mg/l (see Chapter 3; section 3.4.1); so a slight increase in pH might result in manganous carbonate precipitation.

Under the oxic conditions after the initial fast drop, the dissolved manganese concentration decreases continuously indicating that new adsorption sites are created (Figure 5.8b). The creation of the new adsorption sites is most likely due to the oxidation of the adsorbed manganese. This trend of manganese removal under oxic conditions was common to all the filter media and the manganese removal rate per gramme of the media were found to be about $5.2 \times 10^{-4}$ mg per gram media per hour for all the filters A, B and C after the first hour (Figure 5.8b).

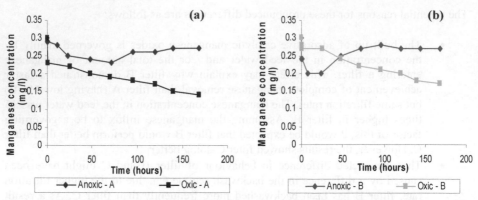

**Figure 5.6** Manganese removal at pH 8 achieved under oxic and anoxic conditions with grains from (a) filter A and (b) filter B.

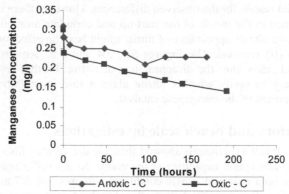

**Figure 5.7** Manganese removal at pH 8 under oxic and anoxic conditions for filter C grains.

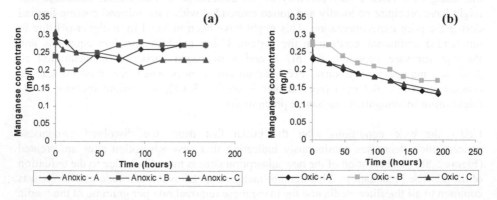

**Figure 5.8** The combined manganese removal graphs at pH 8 for the various filter grains under (a) anoxic and (b) oxic conditions.

Translating the measured removal rate to a filter, assuming a manganese removal rate of $5.2 \times 10^{-4}$ mg per gram media per hour, a filtration rate of 2.8 m/h and 0.6 mg Mn (II)/l in the influent, the following results. The amount of manganese removed in a filter bed with a surface area of 1 $m^2$ and filter height 0.75 m (bulk density 1600 $kg/m^3$), is 0.63 g Mn (II) per hour. The inflow of manganese per hour was however 1.68 gMn (II). So this calculation indicates that the removal rate capacities measured in the laboratory and bench scale tests are too low to explain complete manganese removal obtained with the pilot filters. Possible reasons are:

- the adsorptive catalyst on the top layer from where filter grains were taken has been partially covered with ferric hydroxide that made it partially inactive;
- the mass transfer in a real filter is much higher than in the laboratory and bench scale test;
- the manganese removal capacity has been reduced due to aging in the period between sampling and measuring.

Adsorption isotherms measured under anoxic conditions for six different media, namely: manganese bearing mineral – Aquamandix (AQM), iron oxide coated sand (IOCS), manganese green sand (MGS), laterite, Iron-ore and virgin sand showed that, the manganese bearing mineral (Aquamandix) has by far, the highest adsorption capacity. The virgin sand has almost negligible capacity (Table 5.1; see also Chapter 4, section 4.4.1 for further details). Assuming that the adsorption capacity is the governing factor in developing a catalytic layer on the filter media then, any of the test media is better than the virgin sand. Aquamandix qualifies as a most promising medium to replace sand in situations of slow start –up of manganese removal.

Table 5.1 Freundlich isotherms constant at pH 8

| Filter Media | K | n |
|---|---|---|
| AQM | 875 | 6.6 |
| IOCS | 147 | 3.1 |
| MGS | 64 | 2.8 |
| Laterite | 15.8 | 3.6 |
| Iron-ore | 6.2 | 1.8 |
| Virgin sand | 0.01 | 0.6 |

Unit of K: $[10^{-3}$ mg/g $\times (10^{-3}$ mg/l$)^n]$

## 5.4.3 Development coating and Scanning Electron Microscopy investigation

Photographs of the sand taken from the filters, after 15 weeks of operation, show that the filter materials taken from the top have the darkest color suggesting that they have the highest manganese content. The material from filter B looks darker than the materials from filters A and C. This observation coincides with the fact that this filter had the shortest ripening time (Figure 5.9).

**Figure 5.9** A photograph of the manganese coated sand grains taken fromt the various segments of the filters A, B and C. The samples D0 – 25, D25 – 50 and D50 – 75 represent samples taken from the lowest, middle and top segments respectively of Filter A. Likewisely, E0 – 25, E25 – 50 and E50 – 75 represent sand grains from lowest, middle and top segment of Filter B. The F samples come from the Filter C.

Microscopic pictures taken indicate that coating is not being developed as a uniform layer but is rather developed in the form of spots (Figure 5. 10).

To investigate further the non-uniform development of the coating, a Scanning Electron Microscope (SEM) combined with an EDX probe was used to examine and identify the elemental composition at specific locations on the surface of the grains of the various filter material. The analysis showed that a substantial part of the surface consists of manganese. A small fraction consists of iron. The ratio of manganese to iron ranges from 6.6 to 14.7 (Table 5.2). Surprisingly calcium is present as well; probably the feed was slightly supersaturated with calcium carbonate. Traces of sodium and sulphur were also found.

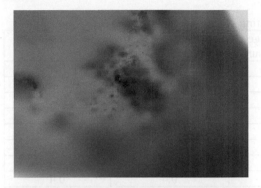

Top layer sand grain of filter A
X 200 magnification

Top layer sand grain of filter B
X 200 magnification

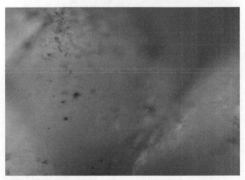

Top layer sand grain of filter C
X 200 magnification

**Figure 5.10** Microscopic pictures of the sand grains taken from the top layer of the various filters.

The spectrum obtained from the examinations at different micro locations on the grain surface of any one of the filter material showed different concentrations of manganese, e.g. Mn concentration obtained through EDX probe analysis of spot A-1 on the surface of a grain from filter A material was much higher than that of spot A-2 of the same material. The irregular coating and development of the catalyst was therefore confirmed by the spectra obtained from the SEM – EDX probe analysis (Figure 5.11) for various locations on the surface of the filter grains taken from filters B and C.

Table 5.2 Elemental composition of coating of filter material

| Elements | Mean percentage element composition at various spots on the surface of filter media grains (%) | | |
|---|---|---|---|
| | Filter A | Filter B | Filter C |
| Mn | 19.0 | 11.9 | 16.2 |
| O (as oxide) | 49.5 | 51.3 | 56.8 |
| Si | 24.4 | 30.8 | 21.9 |
| Fe | 2.2 | 1.8 | 1.1 |
| Ca | 2.9 | 1.7 | 2.0 |
| S | 0.8 | 0.4 | 0.4 |
| Na | 0.5 | 0.4 | 0.4 |

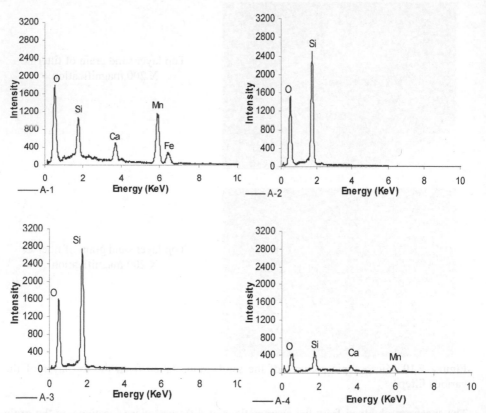

**Figure 5.11** Spectra produced from the Scanning Electron Microscopic and EDX probe analysis at various spots (A-1, A-2, A-3 and A-4) on the surface of the filter A media grains.

## 5.5    Conclusions

In conclusion the following could be deduced:

- The three filters filled with virgin quartz sand developed full manganese removal only after several weeks of operation. The filters fed with water containing 0.6 mg Mn(II)/l and operating at filtration rates of 2.8 and 0.6 m/hr developed complete manganese removal in 79 and 51 days of operation respectively. The reference filter fed with 0.04 mg Mn(II)/l and run at 2.8 m/hr achieved full manganese removal only after 100 days of operation.

- The positive effect of the low rate of filtration is attributed to less frequent backwashing of the filter running at lower filtration rate.

- The long period that the filter fed with low manganese needed to arrive at full manganese removal is attributed to the low concentration of manganese required to develop sufficient adsorptive catalyst surface area.

- Nitrite might have a negative effect on the development of the adsorptive catalyst; however, this has to be verified.

- Adsorption capacity measurements, carried out under oxic conditions on the media of the three filters, taken after 15 weeks of pilot operation, demonstrated a continuous removal of manganese. Conducted batch adsorption experiments are a potentially powerful tool in judging the removal capacity of manganese removing filtering materials. Standardization of the procedure is however, required.

- The measured manganese removal rate capacity of the top layer of the media after 15 weeks of operation, determined in laboratory and bench scale tests, was too low to explain the observed complete removal of manganese in the filters fed with 0.6 mg Mn (II)/l. The manganese adsorptive catalyst on media of the top layers might have been partly covered with ferric hydroxide and / or the catalyst lost a part of its manganese removal capacity due to aging in the period between sampling and measuring.

- Microscopic photographs of the sand grains surface indicate that the catalyst does not develop as a homogenous layer, but develops from specific spots.

- The spectra obtained from Scanning Electron Microscopy investigations on the surface of coated sand grains taken from the top layer of the filters confirmed the non-uniform development of the catalyst in the surface coating.

- Out of six filtration media, a manganese bearing mineral (Aquamandix) demonstrated the highest adsorptive capacity measured under anoxic conditions. As its capacity is much higher than sand, Aquamandix is a potential candidate to substitute (or partially substitute) sand in situations of a slow start-up of manganese removal.

## 5.6 Acknowledgement

The studies in this chapter were done in collaboration with the Kiwa Water Research – Nieuwegein and the Brabant Water Company – 's-Hertogenbosch of the Netherlands. I would like to acknowledge the support given by the two companies and express my appreciation for their assistance both materially and financially.

## 5.7 References

Aqua Techniek, (2007) Aquamandix: http://www.aqua-techniek.com/html/filtermedia.htm.

MWH (2005) Water Treatment Principles and Design. John wiley & sons, Hoboken, NJ, USA.

Stembal, T.; Markic, M.; Briski, F. and Sipos, L. (2004) Rapid start-up of biofilters for removal of ammonium, iron and manganese from groundwater. *Journal of Water Supply: Research and Technology – AQUA* 53.7 (510).

Sharma, S.K. (1999) Adsorptive iron (II) onto filter media. removal from groundwater. *Journal of Water Supply: Research and Technology – AQUA* 48 (3), 84 – 91.

Stumm, W. & Morgan, J.J. (1996) Aquatic Chemistry: Chemical equilibria and rates in natural waters, 3$^{rd}$ edition. Wiley, New York, USA, pp 462.

USEPA (1994) Drinking water criteria document for manganese. Washington, DC, US Environmental Protection Agency, Office of Water.

WHO (2006) Guidelines for Drinking-Water Quality, 3$^{rd}$ edition First Addendum to the third edition Volume 1 recommendations (Vol. 1.) WHO, Geneva, Recommendations.

# CHAPTER SIX

# OPTIMISING THE REMOVAL OF MANGANESE IN UNESCO-IHE ARSENIC REMOVAL FAMILY FILTER TREATING GROUNDWATER WITH HIGH ARSENIC, MANGANESE, AMMONIUM AND IRON

# Abstract

Field studies with UNESCO-IHE arsenic removal family filters showed very good performance under field conditions in rural Bangladesh. However, the presence of manganese in the treated water created an unexpected water quality problem. Most of the filters, filled with iron oxide coated sand (IOCS) demonstrated after several months of operation an increase in manganese concentration in the treated water. These concentrations exceeded largely the concentrations in the groundwater and the WHO guideline value (0.4 mg/l). The goal of this research was to investigate the performance of UNESCO – IHE family filter under laboratory conditions in terms of manganese, arsenic and iron removal efficiency when treating model water with high ammonium concentrations. The hydraulic capacity of the family filter when treating the model water was studied as well. In addition, the effect of introducing polishing layers of Aquamandix and sand in the UNESCO – IHE family filters on manganese, arsenic and iron removal was investigated. With these modified family filters, the ammonium removal phenomenon and the capacity of Aquamandix to remove manganese consistently in the filters without any release of manganese over a period of time have been studied.

In this study, column experiments operated in upflow mode were conducted under laboratory conditions. The filter columns were fed with model water containing high contents of manganese (II) (1 mg/l), arsenite (100 µg/l), arsenate (100 µg/l), iron (II) (5 mg/l) and ammonium (4 mg/l).

Frequent flushing of the IOCS filters was executed to restore the capacity. However, loss of a significant part of the capacity in a period of about two months due to accumulation of iron flocs and biomass could not be prevented.

One filter column, filled with IOCS, has been operated to investigate the removal and release of manganese. A second column was applied to investigate the effect of a promising manganese mineral (Aquamandix) with a high adsorption capacity for manganese. For this purpose a column was filled with IOCS and a layer of Aquamandix (0.5 – 1.0 mm) was provided on the top of the IOCS.

Manganese removal in the filters with IOCS was effective only during a short period of approximately 10 days. Prolonged filter run resulted in manganese release in concentrations exceeding the influent level and the WHO guideline (0.4 mg/l). Replacing the top layer of the filter with fine sand (0.5 – 1.0 mm) and post-treatment with aeration followed by filtration through sand, did not improve manganese removal. The poor removal in these filters is attributed to the relative low pH, covering of the manganese catalyst by ferric hydroxide and biomass and microbial oxidation of ammonium creating anoxic conditions. The release of manganese is due to the anoxic conditions under which $MnO_2$ is being reduced to $Mn^{2+}$.

The filter with Aquamandix on top of IOCS removed manganese effectively without any
release observed. This result indicates that the manganese removal with Aquamandix is
primarily based on adsorption.

A filter filled with IOCS and fed with model water without ammonium demonstrated a
much better removal of manganese, compared to the IOCS filter fed with model water
with ammonium; indicating that anaerobic conditions negatively affect the removal of
manganese.

Nitrite appeared in the effluent of all the IOCS filters in high concentrations exceeding
the WHO guideline (0.2 mg/l). Post-treatment with aeration followed by sand filtration
(0.5 – 1.0 mm) removed this compound adequately. The removal of nitrite is attributed
to biological oxidation by Nitrobacteria.

Arsenic and iron removal in the filter with IOCS was initially satisfactory, however after
two weeks the effectiveness dropped. The filters with Aquamandix or sand as top layer
or post-treatment with aeration followed by sand filter removed adequately arsenic and
iron. This effect is attributed to the removal of small ferric hydroxide flocs, carrying
relatively large amounts of arsenic.

**Key words:** manganese, family filter, arsenic, iron, Aquamandix

# 6.1    Introduction

Over the past six years the UNESCO-IHE has been researching into the use of IOCS in
family filters to treat groundwater with high content of arsenic and iron (Sharma, 2002;
Petrusevski et al, 2003; Salehuddin 2005; Barua, 2006; Petrusevski et al, 2008). The
results obtained from most of the filters installed in Bangladesh in 2004 showed a very
efficient removal of arsenic (about 95 – 99%) and iron from the groundwater through out
the whole testing period of over 30 months without changing the filter media. With
respect to manganese removal, very mixed results were achieved at the 11 testing sites.
Efficient manganese removal was achieved at only two testing sites characterized by low
ammonium concentration in groundwater while the rest suffered from elevated
manganese levels in filtrate. At testing sites with high ammonium in groundwater some
removal of manganese was observed only during the first weeks of operation. With
prolonged operation of the filter, manganese concentrations in the filtrates increased and
at some sites manganese concentration was even higher than its level in the feed water.

At sites with high ammonium concentration, anoxic conditions developed in the filter
bed after some weeks of operation due to biological oxidation of ammonia and possibly
methane present in the feed water. Under such conditions there was shortage of oxygen
to oxidize manganese adsorbed on IOCS during the first weeks (about 3 weeks) of
operation, resulting probably in exhaustion of the manganese adsorption capacity of the
IOCS. Moreover, under anoxic conditions reduction of (partially) oxidized manganese
took place resulting in release of manganese adsorbed or incorporated earlier in the
IOCS matrix (IOCS initially placed in the filters contained approximately 25 mg Mn / g

IOCS. The manganese concentrations in the effluents exceeded in several situations the health based WHO guideline value (0.4 mg/l). Figures 6.1 and 6.2 show the course of manganese in the effluent at two different locations.

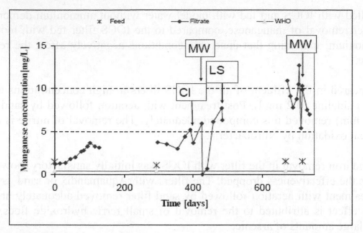

**Figure 6.1** Manganese concentrations in filtrate at location A in Bangladesh. Groundwater: As→ 260 - 280µg/l; Fe → 14.4 - 20.5mg/l; Mn →1.0 - 1.6mg/l; $NH_4^+$ → 5.2 mg/l; $PO_4$ → 1.9mg/l; Total Organic Carbon → 3.0 mg/l, pH 6.9 – 7.0. $HCO_3^-$ → 436 mg/l. MW – Media washing; Cl – Chlorination; LS – addition of limestone layer addition (Source: Barua, 2006).

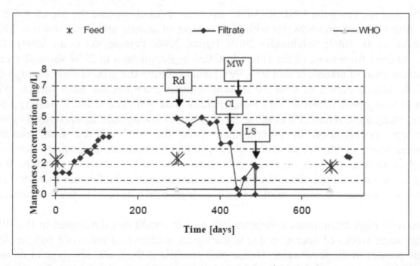

**Figure 6.2** Manganese concentrations in filtrate at location B in Bangladesh. Groundwater: As → 544 – 746 µg/l; Fe → 15.0 - 20.4mg/l; Mn → 1.0 - 1.6mg/l; $NH_4^+$ → 4.3 mg/l; $PO_4$ → 0.4mg/l; Total Organic Carbon → 3.8 mg/l; pH 7.15; $HCO_3^-$ → 491 mg/l: Rd – Daily draining, MW – Media washing, Cl – Chlorination, LS – addition of limestone layer (Source: Barua, 2006).

Chapter 6: Optimising the removal of manganese in UNESCO-IHE arsenic removal family
filter treating groundwater with high arsenic, manganese, ammonium and iron

115

To improve upon the manganese removal and facilitate the hydraulic performance, several measures were implemented including daily and prolonged draining, chlorination, filter media washing, addition of limestone layer and post-sand filtration. These measures provided temporal short-lived improvement in the manganese removal. In most cases, the manganese concentration of the filtrate dropped by 60 – 80 % for at most 1 – 2 weeks (Barua, 2006). It was also observed that nitrite concentration in the filtrate of the filtrates exceeded the WHO Guideline (0.2 mg/l).

Figure 6.2 depicts the beneficial effect of daily draining in combination with chlorination and media washing; however on longer term the effect was not satisfactory. In view of these results, there was the need to develop a method to control the manganese concentration in the effluent of these family filters e.g. to identify new media and measures capable of providing improved filtrate quality for a relatively longer term.

The objectives of this study therefore included investigating the performance of UNESCO – IHE family filter under laboratory conditions in terms of manganese, arsenic and iron removal efficiency when treating model water with high ammonium concentrations. The hydraulic capacity of the UNESCO – IHE family filter when treating the model water with high ammonium, manganese, arsenic and iron concentrations was studied as well. In addition the effect of introducing polishing layers of Aquamandix and sand in the UNESCO – IHE family filters on manganese, arsenic and iron removal was investigated. With these modified filters, the capacity of Aquamandix to remove manganese consistently in the filters without any release of manganese after operating the filters for some time was verified.

To achieve the set objectives two UNESCO – IHE family filters operating in the up-flow mode were used for the investigations. One filter column was filled with IOCS and the other with IOCS and a layer of Aquamandix on the top. These filters were fed with model water having high concentrations of manganese (II), arsenite, arsenate, iron (II) and ammonium. During the operation of the filters, an experiment was done for post-treatment, making use of aeration followed by sand filtration. It was expected that nitrite would be oxidized by bacteria to nitrate, in the post-treatment filter, when sufficient oxygen is introduced.

## 6.2 Theoretical background

### 6.2.1 The role of metallic oxides in arsenic removal

Different treatment technologies like coagulation-sedimentation-filtration, nano-filtration, reverse osmosis, reactor, sub-surface groundwater treatment, adsorption etc. are known schemes which when operated appropriately can remove or reduce adequately arsenic. Adsorption using iron based media like IOCS, granular ferric hydroxide (GFH) etc. has been found to be one of the major promising technologies for removing dissolved arsenic in water because of its simplicity and applicability for both conventional and point-of-use filter systems (Pal, 2001; Petrusevski et al. 2002; Thirunakkarasu et al. 2003; Banerjee and Amy, 2004). In aqueous solutions arsenic occurs predominantly as $H_2AsO_4^-$ and $HAsO_4^{2-}$ ions in the range of pH 5 to 9. Under reducing conditions and pH values below 8, $HAsO_2$ $_{(aq)}$ or $H_3AsO_3$ mainly occurs (i.e. arsenic exists in the +3 state) (Hem, 1977). Therefore in the pH range (i.e. $6 - 8.0$) normally associated with groundwater these arsenic ions can be found.

Iron and manganese oxyhydroxides play major and additional roles respectively in arsenic removal from groundwater. When groundwaters with pH values within the range $6 - 8$ and high content of manganese, iron and arsenic, are aerated, the dissolved Fe(II) and As (III) get oxidize to Fe (III) and As (V) respectively. Oxidation with As(III) with oxygen is, however, slow. The As (V) (i.e. arsenate) thus formed gets removed through adsorption / co-precipitation by the hydrolyzed Fe (III) ions. Oxidation of dissolved manganese through aeration at pH values below 9 is slow and therefore dissolved manganese in groundwater may not contribute much to the arsenic removal. Manganese oxides coatings formed on the surface of the filter media, most likely contribute substantially to the oxidation of As (III). Borho and Wilderer, 1996 demonstrated $MnO_2$ to be a most promising oxidants for As (III) oxidation.

Arsenate is much more effectively removed than arsenite since arsenic (V) exists as monovalent or divalent anions ($H_2AsO_4^-$, $HAsO_4^{2-}$) whereas arsenic (III) occurs in the neutral form ($H_3AsO_3$) (Hug and Leupin, 2003). In view of this, to enhance arsenic removal arsenite oxidation is facilitated through either the addition of an oxidizing reagent e.g. $Cl_2$, $KMnO_4$, $O_3$, or the application of catalytic oxides like manganese oxides (Lee et al. 2003). In another development, $\delta$-$MnO_2$ has been reported to enhance faster oxidation of arsenic (III) than the other manganese dioxides minerals (Driehaus et al. 1995).

The oxidation of arsenite to arsenate by manganese dioxide might result in the release of $Mn^{2+}$ ions (equation 6.1). For 1 mg As (III) oxidized, 0.73 mg of Mn (IV) is reduced. The $Mn^{2+}$ ions produced adsorb onto the manganese dioxide and give it a positive surface charge that facilitate the removal of the arsenate present initially and produced after the arsenite oxidation (Bajpai and Chaudhuri, 1999).

$$H_3AsO_3 + MnO_2 \rightarrow HAsO_4^{2-} + Mn^{2+} + H_2O \qquad (6.1)$$

During filtration in groundwater treatment, dissolved arsenic get co-precitated and / or adsorbed mostly onto $Fe(OH)_3$ and to some extent on $MnO_2$. Bissen and Frimmel, (2003), observed that arsenic ions are adsorbed much more onto $Fe(OH)_3$ particles than $MnO_2$ particles, when aerated groundwater was infiltrated in the subsurface with the intention of removing arsenic from an arsenic contaminated groundwater.

## 6.2.2 Influence of ammonium and methane on manganese, arsenic and iron removal

One of the characteristics of anaerobic groundwater is the frequent presence of ammonium and methane etc. The quantities of these two compounds depend on the presence of organic matter (e.g. peat) in the soil.

*Ammonium*

Under aerobic conditions ammonium is oxidized by bacteria to nitrite and subsequently nitrate. The term nitrification is generally used to describe the biological oxidation of ammonium nitrogen ($NH_4^+$-N) to nitrite ($NO_2^-$-N) and nitrate ($NO_3^-$-N). The nitrification is a two-step process catalysed through the metabolic activity of two groups of autotrophic bacteria (i.e. their carbon source is inorganic). The two groups include the *Nitroso*-bacteria and *Nitro*-bacteria. Commonly noted for nitrification are the autotrophic bacteria *Nitrosomonas* and *Nitrobacter* which oxidize ammonia to nitrite and then to nitrate respectively. Other autotrophic bacteria genera capable of obtaining energy from the oxidation of ammonia to nitrite have been reported. These include *Nitrosococcus*, *Nitrosospira*, *Nitrosolobus* and *Nitrosorobrio* (Painter, 1970). Also reported are autrotrophs capable of transforming nitrite to nitrate other than *Nitrobacter*; including *Nitrococcus*, *Nitrospina*, *Nitroeystis* etc. (Teske et al, 1994; Wagner et al, 1995). The nitrifying bacteria are aerobes and chemolithoautotrophs. The autotrophic nitrification proceeds optimally within a temperature range of $25 - 35°C$ and pH $7 - 8$. The two step oxidation of ammonium to nitrate can be represented as follows:

$$2NH_4^+ + 3O_2 \rightarrow 2\,NO_2^- + 4H^+ + 2H_2O \qquad \text{(Nitroso-bacteria)} \qquad (6.2)$$

$$2NO_2^- + O_2 \rightarrow 2NO_3^- \qquad \text{(Nitro-bacteria)} \qquad (6.3)$$

Total oxidation

$$NH_4^+ + 2O_2 \rightarrow NO_3^- + 2H^+ + H_2O \qquad (6.4)$$

From the equation of the total oxidation reaction (6.4), the oxygen required to completely oxidize 1 mg of ammonium is 3.56 mg with 2.67 mg $O_2$ /mg used for nitrite production and 0.35 mg $O_2$ used for every mg of nitrite oxidized to nitrate (Tchobanoglous et al, 2003). As the ammonium conversion proceeds, part of the ammonium is assimilated into formation of cell tissue as follows:

$$4\,CO_2 + HCO_3^- + NH_4^+ + H_2O \rightarrow C_5H_7O_2N + 5O_2 \qquad (6.5)$$

where $C_5H_7O_2N$ is used to represent cell tissue.

In groundwater treatment e.g. applying aeration followed by rapid sand filtration, the presence of ammonium might result in anaerobic conditions. When 10 $mgO_2/l$ is present after aeration, concentrations exceeding 2.7 mg $NH_4^+/l$ will result in the creation of anaerobic conditions within a main or a part of the filters. These anaerobic conditions will prevent oxidation of the adsorbed manganese (II) and might cause the reduction of manganese oxides, present on the surface of filter media, resulting in disturbed manganese removal and / or release of manganese. In addition the oxidation of $As^{3+}$ will be hindered as well. Usually the oxidation of $Fe^{2+}$ is less affected, since the rate of oxidation of $Fe^{2+}$ is rather high.

A second group of autotrophic nitrifying bacteria (e.g. *Nitrosomonas europaea*) exist that are capable of using nitrite to oxidize ammonia under anaerobic conditions. In other words, as the ammonia is oxidized the nitrite gets reduced to produce nitrogen gas. In the presence of dissolved oxygen, the *Nitrosomonas europaea*, oxidizes the ammonia with oxygen as the electron acceptor (Bock et al. 1995).

Another phenomenon, discovered in the mid-1990's explains the conversion of ammonium into nitrogen by yet, another different group of bacteria referred to as the Anammox bacteria. The Anammox bacteria oxidize ammonia using nitrite as the oxidant under anaerobic condition but can not use oxygen for ammonia oxidation (Jetten et al 1999, Gable and Fox, 2003). The Anamox bacteria have been successfully applied in waste water treatment but its presence in groundwater treatment has not been reported. In all these nitrification reactions, some nitrate may be generated as one of the end products.

*Methane*
Methane is commonly associated with ammonium in groundwater. Concentrations up to 40 mg/l are reported in The Netherlands. Under aerobic conditions methane is easily oxidized by bacteria forming large quantities of biomass.

$$CH_4 + 2O_2 \rightarrow CO_2 + 2H_2O \qquad\qquad (6.6)$$

According equation 6.6, 1 mg $CH_4$ needs 4 mg $O_2$. As a consequence for groundwater containing 10 $mgO_2$/l after aeration, filters are able to remove a maximum of 2.5 mg $CH_4$/l. Higher concentrations will result in anaerobic conditions and that is why intensive aeration (e.g. with plate aerators or aeration towers) is frequently applied in practice to reduce methane to acceptable levels before rapid sand filtration. Anaerobic conditions in filters caused by methane oxidation will disturb manganese removal and could result in release of adsorbed manganese and also hinder the oxidation of As (III) and Fe(II).

## 6.3    Family filter for arsenic removal

Following several trials in the UNESCO-IHE laboratories, prototypes of the Family Filter were installed in Bangladesh in 2004 – 2006 and tested in the field over the period. Several modifications made to the original design with the intent of improving upon the hydraulic performance and the filtrate quality resulted in the manufacture of the third generation Family filter as shown in Figure 6.3. The UNESCO-IHE family filter comprise a raw water bucket, a 150 mm diameter PVC filter column with a 50 cm IOCS (particle size: 2 – 4 mm) bed placed on top of a 10 cm supporting media of IOCP (iron oxide coated pumice; particle size: 5 – 15 mm). Above the IOCS bed is placed a small polishing layer normally about 5 cm depth of fine IOCS. A number of control off / on valves are provided to allow the operation and draining of the filter (Figure 6.4).

Aeration of the anaerobic ground water is achieved during the filling of the raw water bucket in the residence. Due to the introduced oxygen, the following reactions are assumed to take place in the filter: i) adsorption of the iron (II), ii) oxidation of the adsorbed and dissolved iron (II), followed by bio-oxidation of the methane and ammonium and iii) adsorption of manganese (II) followed by oxidation.

**Figure 6.3** A third generation UNESCO-IHE Family Filter installed in a village in
Bangladesh.

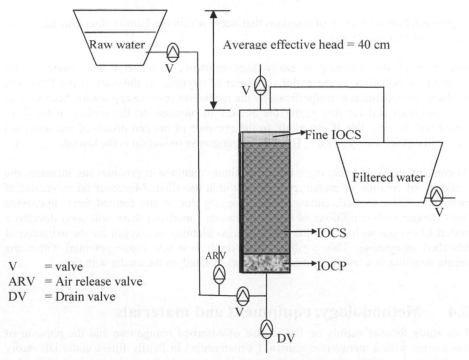

**Figure 6.4** Schematic diagram of the UNESCO – IHE Family filter.

The assumed reactions are in accordance with what has been reported in literature and given in Figure 6.5 (Stumn and Morgan, 1996).

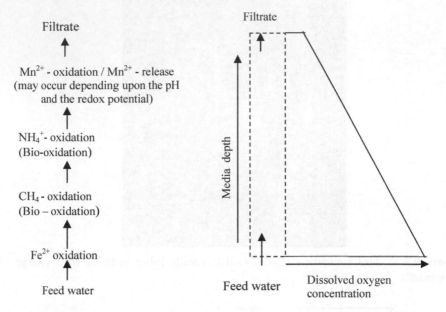

**Figure 6.5** Probable series of reactions that occur within the Family filter (Adapted from Stumn and Morgan, 1996).

Oxidation of the methane is easily accomplished by bacteria and together with ammonium oxidation, is the major consumer of oxygen. At the start of the filter run, methane bio-oxidation is insignificant as the media has prior to application been kept in dry conditions and for that matter the amount of biomass on the surface of the filter media will be very low if any at all. In the presence of oxygen dissolved and adsorbed $Fe^{2+}$ will oxidize easily to $Fe^{3+}$. The rate of manganese oxidation is the lowest.

At concentrations of a few mg/l, the ammonium / methane in groundwater increases the tendency of creation of anoxic conditions within the filter. Moreover large amount of biomass will be formed, enhancing the clogging due to the formed ferric hydroxide flocs. Under such conditions of high ammonium / methane there will soon develop a deficit of oxygen within the filter and thus unavailability of oxygen for the oxidation of adsorbed manganese. This condition may result in a low redox potential within the media resulting in a release of manganese accumulated on the media with time.

# 6.4    Methodology, equipment and materials

This study focuses mainly on the release of adsorbed manganese and the removal of manganese with a manganese mineral (Aquamandix) in family filters under laboratory conditions. In addition the improvement of the removal of arsenic, iron and nitrite has

Chapter 6: Optimising the removal of manganese in UNESCO-IHE arsenic removal family
filter treating groundwater with high arsenic, manganese, ammonium and iron

121

been investigated. For this purpose, the experiments were organized into 3 major parts
(part 1, 2 and 3) with a total four different column experiments conducted under
laboratory conditions.

## 6.4.1 Column experiment

During the first part of the column experimentation (i.e. part 1), two UNESCO-IHE
family filters - one filled with IOCS (filter 1) and the other with IOCS and Aquamandix
on top (filter 2); were installed and ran with model water containing Mn(II) (1 mg/l),
Fe(II)(5 mg/l), As(III) (100 µg/l), As(V) (100 µg/l) and $NH_4^+$(4 mg/l) to investigate the
removal and release of manganese over a period of time (Figure 6.6). Figure 6.6 shows
the schematic diagram of the set up of the experiments with filters 1, 2 and 3.

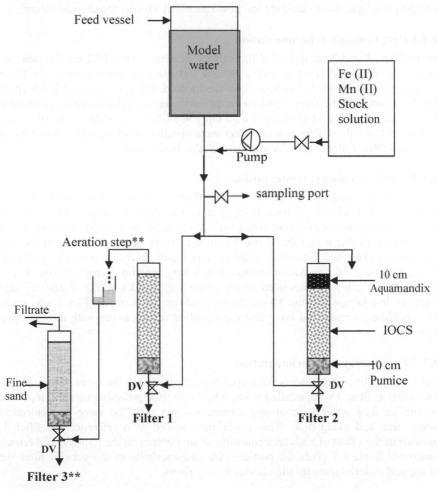

**Figure 6.6** Schematic diagram of filter set-up; filter 3 and the aeration step were
installed after one month of operation.

After one month of operation, part 2 of the column experiment was initiated by installing filter 3 containing virgin sand. The filter 3 was fed with aerated effluent of the filter 1 (Figure 6.6) to investigate the effect of aeration followed by sand filtration, as a post-treatment, on nitrite and ammonium concentrations. After two months, filter 1 was modified by placing a layer of fine sand on top of it, to serve as a polishing layer to remove escaping flocs.

For part 3, filter 4 was installed equipped with IOCS and a polishing sand layer on top of it. The filter 4 was run with model water with the same concentration of arsenic, iron and manganese but without ammonium to act as a reference filter. For the outlet of the filters, orifice of size 0.6 mm was used. A flow rate of 2.1 l/h corresponding to a filtration rate of 0.25 m/h was used for all the filters. Using the 0.6 mm orifice, the corresponding head determined for the flow rate of 2.1 l/h was found to be 40 cm.

### 6.4.1.1 Part 1 column experimentation

Family filters 1 and 2 comprised a 70 cm PVC column with a 10.2 cm diameter and a bed depth of 63 cm. A layer of coarse pumice was placed at the bottom of the filters as the supporting media for the various filter media used. Filter 1 contained IOCS (particle size: 2 – 4 mm) and the filter 2 had 10 cm polishing layer of Aquamandix (particle size: 0.5 – 1.0 mm) placed on top of about a 53 cm IOCS filter bed (Table 6.3). All necessary accessories like valves, fittings and covers were installed allowing a free board of 5 to 8 cm in each filter. Filters 1 and 2 were fed with the model water.

### 6.4.1.2 Part 2 column experimentation

After operating filters 1 and 2 for one month, filter 3 together with an aeration step was installed and fed with the filtrate from flter 1. The filter 3 comprised a 35 cm PVC column with a diameter of 10 cm filled with 20 cm depth of washed sand (particle size: 0.5 – 1.0 mm) (Table 6.3). Filter 3 had 10 cm deep layer of coarse pumice at the bottom as supporting layer and 5 cm free board on top. Upon observing positive results of the filter 3, regarding the removal of arsenic, a polishing sand layer (particle size: 0.5 – 1.0 mm) of 10cm depth was introduced on top of the IOCS media in filter 1 after 57 days of operation. In addition, the top 10 cm segment of the IOCS in the filter 1 was replaced with a polishing virgin sand layer and the modified filter was run with the same model water.

### 6.4.1.3 Part 3 column experimentation

After two months of continuous filter operation, filter 4, with the same designed column dimensions as filter 1 was installed using IOCS with sand polishing layer on top. Filter 4 was run on feed water without any ammonium but with the same concentration of arsenic, iron and manganese. This experiment served as a reference for filter 1, to demonstrate the effect of high concentration of ammonium on the removal and release of manganese. Table 6.1 gives the particle size characteristics of the various filter media and support material used for the various family filters.

**Figure 6.7** A photograph of (a) the installed UNESCO-IHE Family filters in operation
(b) Aquamandix media (c) IOCS grains (d) Polishing Sand (e) Pumice grains.

**Table 6.1** Particle size characteristics for the various filter and supporting media.

| Type of media | $d_{10}$ (mm) | $d_{60}$(mm) | $d_s$ (mm) | $U_c$ |
|---|---|---|---|---|
| Pumice | 6.50 | 9.50 | 7.80 | 1.46 |
| IOCS | 2.82 | 3.50 | 3.22 | 1.24 |
| Sand | 0.68 | 0.78 | 0.76 | 1.14 |
| Aquamandix | 0.65 | 0.76 | 0.74 | 1.18 |

$U_c$ = Uniformity coefficient,      $d_s$ = Specific diameter

Model water mirroring typical groundwater composition Terai part of Nepal was prepared from Delft tap water by increasing the $HCO_3^-$ concerntration to 4.5 mmol/l (i.e. 275 mg$HCO_3^-$/l) with Analar grade sodium hydrogen carbonate and decreasing the pH to 6.8 – 7.0. Other Analar grade reagents used in the preparation of the model water and their respective concentrations have been detailed out in Table 6.2. The other reagents used for the model water preparation included manganese (II) sulphate, arsenic (III) oxide and arsenic (IV) oxide, ammonium chloride and hydrochloric acid for the manganese, arsenic, ammonium dosing and pH adjustments respectively.

**Table 6.2** Water composition of the model water .

| Parameter | Model Water (mg/l) | Parameter | Model Water (mg/l) |
|---|---|---|---|
| pH | 6.8 | Oxygen ($O_2$) | 10.5 |
| Hydrogen carbonate ($HCO_3^-$) | 275 | Calcium ($Ca^{2+}$) | 53 |
| Iron ($Fe^{2+}$) | 5 | Silicate ($SiO_3^{2-}$) | 1.5 |
| Manganese ($Mn^{2+}$) | 1 | Sulphate ($SO_4^{2-}$) | 81 |
| Ammonium ($NH_4^+$) | 4 | Phosphate ($PO_4^{3-}$) | 0.03 |
| Arsenic (III) | 0.1 | Chloride ($Cl^-$) | 61 |
| Arsenic (V) | 0.1 | Sodium ($Na^+$) | 40 |
| Temperature (°C) | 16 | Total Organic Carbon | 1.2 |

### 6.4.1.4 Sampling criteria and water quality analyses

During the filter runs, the feed water and the filtrate of the filters were sampled periodically, acidified with 2 N HCl and subsequently analyzed for total manganese, arsenic, iron, ammonium, nitrate, nitrite, pH and dissolved oxygen. The analysis was done according to the Dutch Standard Method NEN 6457. For the total arsenic concentration determination, the atomic absorption spectrometer – Thermo Elemental Solaar MQZe – GF 95 (AAS-GF) equipped with graphite furnace and an autosampler with arsenic detection limit of 2 µg/l was used. The total dissolved iron and manganese determination for the samples were conducted using the ICP-Perkin Elmer Optima 3000 and the Perkin Elmer Atomic Absorption spectrometer 3110 with detection limits of 0.02 mg/L for manganese and 0.05 for iron. The $NH_4^+$ - N and $NO_2^-$ -N in the samples were determined using the Perkin Elmer lambda 20 spectrophotometer. The pH and dissolved oxygen were determined using WTW 323 portable pH and WTW oximeters respectively.

**Table 6.3** Summary of the media composition and experimental treatment of the various filters.

| Filter | Media | Media particle size (mm) | Treatment |
|---|---|---|---|
| Filter 1 | IOCS | 2 - 4 | model water |
| Filter 1 modified | IOCS | 2 – 4 | |
| | Sand on top | 0.5 – 1.0 | |
| Filter 2 | IOCS | 2 - 4 | model water |
| | Aquamadix on top | 0.5 -1.0 | |
| Filter 3 | Sand | 0.5 – 1.0 | aerated effluent filter 1 |
| Filter 4 | IOCS | 2 - 4 | model water without |
| | Sand on top | 0.5 – 1.0 | ammonium |

In order to determine the possible presence of particulates (micro-flocs, carrying iron, manganese and arsenic), in the filtrate, the sampled filtrates were filtered through 0.45 μm membrane filters. The flow rate of the feed water was monitored periodically and if required adjusted.

### 6.4.1.5 Filter operations

The filters operating in the upflow mode were drained averagely twice a week to maintain the hydraulic capacity of the units through removal of iron deposits from the filter media. To accomplish the draining process the feed water valves were closed and the draining valves opened and the filters flushed at a high discharge rate. Aside cleaning of the filter units, the draining process facilitates oxidation of adsorbed iron and and to limited extent, manganese. In the course of the experiments, it was realized that the filter 2 (i.e. the filter with the Aquamandix) had to be drained daily due to the rapid decrease of capacity.

## 6.5    Results and discussions

## 6.5.1 Column experiments

### 6.5.1.1 Filter capacity and regular draining

At the initial stages of the filter runs, the filters 1(with IOCS) and 2 (with IOCS and Aquamandix on the top) were fed with model water containing high ammonium, iron, manganese and arsenic contents at the rate of 2.1 L/h. From the regular monitoring of the capacities of the filters, it was observed that a marked decrease in the capacities of the filters 1 and 2 (i.e. 35% and 60% decrease for filter 1 and 2 respectively) occurred after one week of operation (Figures 6.8 and 6.9). In order to restore the capacity, intermittent draining of the filters was conducted after the first week and that led to about 80 – 100% recovery of the initial capacity; however the recovery was not sustainable. The extent of recovery dropped continuously over the period of the experimentation (about 60 days). The recovered capacity of the filters 1 and 2 experienced an average daily reduction of their initial capacities of 0.48% and 0.13% respectively.

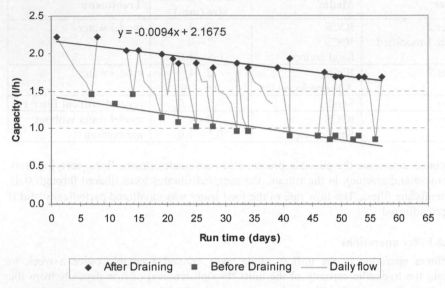

**Figure 6.8** Capacity of filter 1 with IOCS as a function of run time and filter draining (Average $V_f$ – 0.25 m/h; Model water: As (III) – 100 µg/l; As (V) – 100 µg/l; $Fe^{2+}$ – 5 ± 0.5 mg/l; $Mn^{2+}$ = 1± 0.5 mg/l; $HCO_3^-$ = 275 mg/l; $NH_4^+$ = 4 mg/l; pH = 6.8).

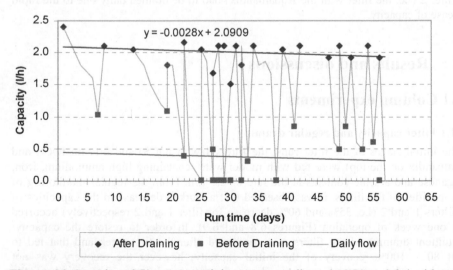

**Figure 6.9** Capacity of filter 2 (containing Aquamandix and IOCS and fed with model water) as a function of run time and filter draining.

The Figures 6.8 and 6.9 show that, in between any two draining sessions the drop in capacity for the filter 2 was more than twice that of Filter 1 possibly due to the smaller particle size of the Aquamandix polishing layer and the eventual built up of a higher resistance. The resistances in both filters are due to the formation and accumulation of iron flocs and biomass development.

### 6.5.1.2 Performance of filters in removal of manganese, ammonium, arsenic and iron

#### 6.5.1.2.1 Manganese removal and the influence of ammonium

At the start of the column experiments, filters 1 and 2 were fed with model water with high contents of manganese, iron, arsenic and ammonium. After an initial satisfactory manganese removal to levels below 0.4 mg/l (The WHO health based guideline), the manganese content in the filtrate of filter 1 gradually shot up to levels similar to that of the feedwater (Figure 6.10). After one month of operation, the filtrate manganese concentration of filter 1 was above that of the feedwater; obviously manganese was leaching out from the IOCS filter media.

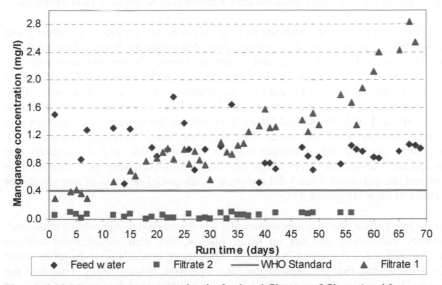

**Figure 6.10** Manganese concentration in feed and filtrates of filters 1 and 2

In filter 2, manganese was consistently removed below the 0.4 mg/L level throughout the experimental period; showcasing a manganese removal efficiency of 90 – 100% (Figure 6.10). The high manganese removal capacity of the Aquamandix is the predominant factor in the efficient manganese removal. The dissolved oxygen monitoring conducted on the filtrates of the two filters (1 and 2) showed the prevalence of anoxic conditions after ripening period of a few weeks (Table 6.4). The anoxic conditions within the columns were expected as the oxidation of iron and ammonium proceeded. Concurrently pH within the filter 2 increased slightly by about 0.02 - 0.04 units. Under the normal circumstances a drop in pH would have been expected as adsorption and oxidation of iron and ammonium proceed with the production of H$^+$ ions. It is therefore not unlikely that the pH has been buffered by the IOCS due to presence of calcium carbonate in its coating. Ammonia removal in the filters increased gradually till it peaked at about 35 days when all ammonia present in feed water was consistently removed (about 80 – 100% removal) (Figure 6.13).

**Table 6.4** The range of dissolved oxygen values and pH of feed water and filtrates.

| Filter | Dissolved Oxygen (DO) (mgO$_2$/l) | | pH | |
|---|---|---|---|---|
| | Feed water | Filtrate | Feed water | Filtrate |
| Filter 1 | 7.21-10.4 | 0.15 - 1.0 | 6.5-7.2 | 6.7-7.2 |
| Filter 2 | 7.21-10.4 | 0.1 - 0.5 | 6.5-7.2 | 6.9-7.4 |
| Filter 3** | 4.9-8.3 | 3.0 - 6.3 | 7.1-7.9 | 6.9-7.1 |

** The results of the third filter installed after 1 month (i.e. 30 days) of filter operation

The results shown in Table 6.4 indicate that anoxic conditions prevail in the filters 1 and 2 meaning that oxygen consuming reactions proceeded within the filters. These reactions include bio-oxidation of ammonium and oxidation of iron. The low pH range indicates that, no oxidation of manganese in the water phase is to be expected. The prevailing conditions are not favourable for catalytic manganese removal as well because of the low pH and absence / low levels of oxygen. In the filter 3, conditions were different because it was fed with the aerated filtrate from filter 1, the amount of ammonium and iron in its influent were far lower than those of filters 1 and 2 (Figure 6.11). Consequently consumption of oxygen was on the lower side and the aeration step introduced led to an increased pH within the filter 3. However, after more than 35 days of filter run, the polishing filter 3 did not show a substantial removal of manganese for the effluent of filter 1(Figure 6.11).

To investigate further, the role ammonium plays in the leaching / removal of mangnese, a reference experiment was set up using filter 4, fed with model water without ammonium and approximately 1.0 mg Mn/l. Manganese was relatively better removed by IOCS during the first 28 days of the filter operation (Figure 6.12). Therefore the manganese level in the filtrate remained below the WHO guideline value of 0.4 mg/l over that period however, the level of manganese in the filtrate rose gradually afterwards with time (Figure 6.12).

Chapter 6: Optimising the removal of manganese in UNESCO-IHE arsenic removal family filter treating groundwater with high arsenic, manganese, ammonium and iron

129

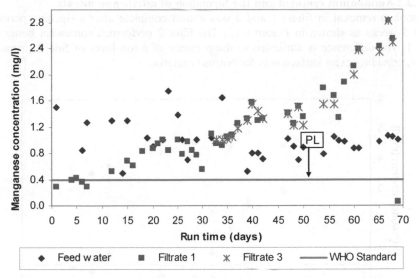

**Figure 6.11** Manganese concentrations in feed and filtrate in filter 1 and filter 3
PL is the moment when polishing sand layer of depth 10 cm was introduced.

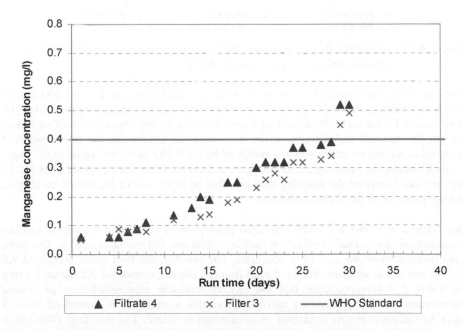

**Figure 6.12** Manganese removal performance of filter 3, fed with aerated effluent of filter 1 and filter 4 fed with model water without ammonium.

**6.5.1.2.2 Ammonium removal and the formation of nitrite and nitrate**

Ammonium removal in filters 1 and 2 was almost complete after a ripening period of about 5 weeks as shown in Figure 6.13. The filter 2 performed somewhat better than filter 1. The difference is attributed to the presence of a top layer of fine Aquamandix grains, providing extra surface area for Nitroso bacteria.

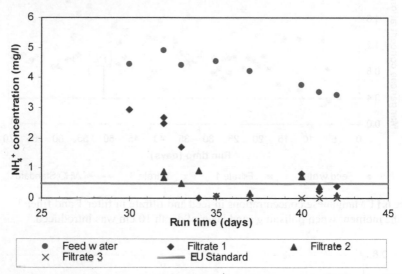

**Figure 6.13** Concentration of ammonium ($NH_4^+$) in feed water and effluent of filters 1 and 2. Filter 3 was fed with aerated effluent of filter 1.

Simultaneously, the nitrite monitoring conducted on the filtrates of filters 1 and 2 after the 35 days period showed high levels of nitrite (up to 2.15 mg/L) (Figure 6.14). The main reason for the high nitrite content would probably be the unavailbility of sufficient oxygen for survival and activity of the nitro-bacteria. Eventually the nitrite was not oxidized to nitrate; so nitrite was not removed below WHO guideline value of 0.2 mg/l. Co-incidentally, the leaching out of manganese from the filter 1 also started after the 35 days period. Therefore the leaching may possibly be as a result of the anoxic conditions and the presence of the nitrite molecules.

The Table 2 in Annex 6.1 gives a detailed calculation of actual and theoretical oxygen consumptions for oxidation of ammonium to nitrite and nitrate. Data used in this table were taken after 40 days of filter operation. On that day, the feed water contained 9.5 $mgO_2/l$ and it was assumed that the 5 mg/l iron (II) present, consumed 0.7 mg $O_2/l$. From the Table 2 it becomes clear that the oxygen available was insufficient to oxidize ammonium completely to nitrite and moreover that a total deficit existed of about 3 $mgO_2/l$ to allow complete oxidation of ammonium to nitrate. The polishing filter (filter 3) removed all residual ammonium from the effluent of the filter 1 and converted all the existing and formed nitrite into nitrate, obviously the aeration step provided the needed oxygen to oxidise the ammonium completely to nitrate (Figure 6.13).

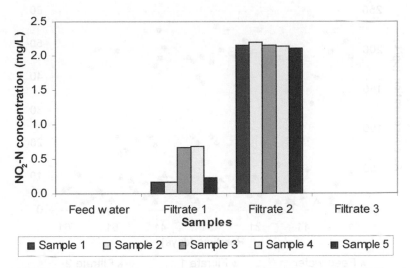

**Figure 6.14** Concentration of nitrite in feed water and effluent of the different filters after the ripening period.

### 6.5.1.2.3 Arsenic removal

The arsenic removal efficiency of filter 1 consistently ranged from 95 – 99% during the first week of operation however this removal efficiency dropped to about 75 – 95% for the following 7 weeks (Figure 6.15). The arsenic concentration in the filtrate after a period of two weeks was predominantly above the WHO guideline value of 10µg/l. The reduction in arsenic removal efficiency was contrary to the consistent 97 – 99 % arsenic removal efficiency reported by Salehuddin (2005) and Barua (2006) who had used the UNESCO – IHE filters for up to 30 months in field studies in Bangladesh.

This discrepancy raised the suspicion that perhaps arsenic was escaping attached to particulates of either iron flocs or manganese within the filter 1 of the current experiment. To investigate the discrepancy, the unacidified filtrates of filters 1 & 2 were were immediately after sampling filtered through 0.45 µm membrane filter. The Figure 6.16 shows the arsenic concentrations after filtration through the 0.45 µm membrane filter being significantly lower in the effluent of filter 1 and almost nil for filter 2 confirming that arsenic was escaping adsorbed and / or co-precipitated on iron flocs.

To curtail the trend of decreasing arsenic removal efficiency, filter 3 was installed after four weeks of filter operation using washed sand (particle size: 0.5 – 1.0 mm) and fed with aerated filtrate from filter 1. The polishing filter removed effectively arsenic to the level below WHO standard of 10µg/l (Figure 6.17). Equally effective was the removal of iron from the filtrate fed to the filter 4 (Figure 6.21).

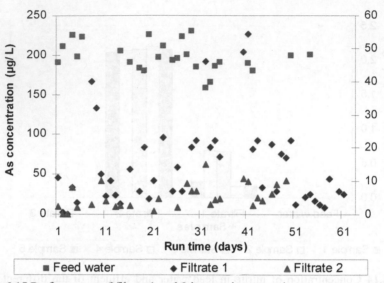

**Figure 6.15** Performance of filters 1 and 2 in arsenic removal.
The left and right hand side vertical axes present scales meant for arsenic concentration in the feedwater and filtrate respectively.

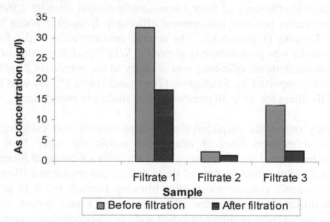

**Figure 6.16** Arsenic concentration in filtrate before and after filtration through 0.45 μm filter membrane.

This high removal efficiency of the filter 3 (and filter 4) was likely due to the following reasons:

- Adsorption of As onto the fine particles of oxidised iron in filtrate 1. An effective removal of these particles including the attached arsenic by the fine sand.

- The aeration might have enhanced the oxidation of As (III) to As (V), and the latter is better adsorbed onto ferric hydroxides than the As (III). However, there are strong indications that oxidation of As (III) needs a catalyst in the form of iron (II) and / or manganese (IV). Oxidation of As (III) to As (V) by dissolved oxygen proceeds at a slow rate (Driehaus et al, 1995).

As performance of the filter 3 was very efficient on arsenic removal, it was decided to place a polishing layer of fine sand layer on top of IOCS in filter 1. So after 51 days of operation, a polishing layer of 10 cm thick fine sand (particle size: 0.5 mm - 1.0 mm) was placed at the top of the IOCS in filter 1. Upon addition of this layer, the arsenic removal efficiency increased significantly up to 97.5% (Figure 6.17). This explains the difference in arsenic removal obtained during the field work in Bangladesh and that from the laboratory test since in Bangladesh such a polishing layer (of fine IOCS) was placed at the top of the IOCS layer.

The arsenic removal efficiency of filter 2 was consistently at 94 – 99% throughout the experimental period (Figure 6.15). The arsenic concentration of the filtrate of filter 2 was almost always ≤ 10µg/l. The reasons for the better removal capacity of the filter 2 are attributed to the polishing layer of Aquamandix and effective removal of iron flocs with incorporated arsenic. It can not be excluded that arsenic is adsorbed on the Aquamandix. An indication that adsorption did occur is that the effluent of filter 1, after filtration through a membrane filter (0.45 µm) contained about 15 µg/l, while the unfiltered effluent of filter 2 had very low arsenic concentration (Figure 6.16).

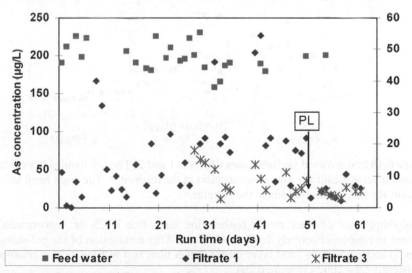

**Figure 6.17** Performance of filter 1 and filter 3 in arsenic removal. The filter 3 was fed with effluent of filter 1. PL = moment when polishing sand layer of depth 10 cm was introduced on filter 1. The left and right hand side vertical axes present scales meant for arsenic concentration in the feedwater and filtrate respectively.

In the reference filter 4 fed with model water without ammonium, arsenic removal remained consistently at about 99% throughout the experimental period. In addition to effective arsenic adsorption on IOCS, the high arsenic removal efficiency could be explained with the very effective removal of iron micro-flocs with attached arsenic in the top polishing sand layer placed at the top of the IOCS.

### 6.5.1.2.4 Iron removal

For the first 3 weeks of filter operation, the iron content of the filtrate of filter 1 remained predominantly below 0.3 mg/l (the concentration value beyond which consumers normally complain about the aesthetic nature of the water). Throughout the period, more than 90% iron removal efficiency was recorded however this efficiency dropped to 70 – 90% for the remaining experimental period (Figure 6.18). About 30% of the iron in the filtrates of filter 1 existed as iron flocs that passed through the filters (Figure 6.19). This confirms the assumption that the increase in arsenic concentration in the filtrate of filter 1 is because of escaping iron micro-flocs with attached arsenic ions.

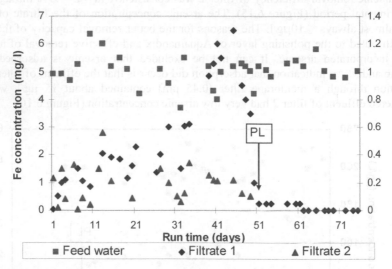

**Figure 6.18** Iron removal performances of filters 1 and 2. The left hand side vertical axis presents a scale meant for iron concentration in the feedwater. The right hand side scale is meant for iron concentrations in the filtrates.

A polishing layer of filter media (either fine sand, fine IOCS or Aquamandix) was required to remove effectively ferric oxide flocs. After installation of the polishing filter (filter 3) and a polishing sand layer on the IOCS filter bed in filter 1, iron removal was consistently 99 – 100 % (Figure 6.20).

The filter 2 exhibited a better iron removal efficiency of 95 – 100% through out the experimental period (Figure 6.18). The high Iron adsorption capacity could be attributed to the ferric oxides of both IOCS and the Aquamandix and the aluminium oxides of the Aquamandix. On Figure 6.20, the installation of filter 3 and introduction of a polishing

layer in the filter 1 resulted in the removal of oxidised iron flocs predominantly through
straining action. In the event of these flocs having likely adsorbed arsenic ions, the
arsenic concentration in the filtrates of the filters 1 and 3 was expected to drop
remarkedly (Figure 6.17). The same concentration of arsenic in filtrates of both filter 1
(after polishing layer introduction) and filter 3 as observed on Figure 6.17 strongly
proved the straining action of fine sand.

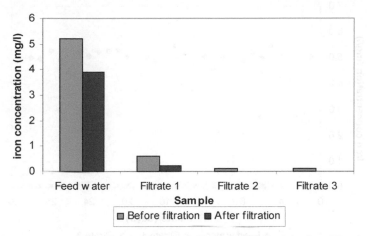

**Figure 6.19** Iron concentration in feed and filtrate before and after filtration through
0.45 μm filter.

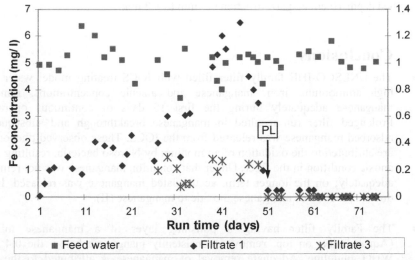

**Figure 6.20** Performance of filters 1 and 3 in iron removal. Filter 3 was fed with aerated
effluent of filter 1. The left and right hand side vertical axes present scales meant for
iron concentrations in the feedwater and filtrate respectively.

Iron removal efficiency was nearly 100% right from the beginning of the reference filter experimentation and remained consistently at that level of efficiency till the end (Figure 6.21). In the absence of the ammonia, sufficient oxygen was available for the oxidation of adsorbed iron (II) and probably the oxidation of As (III) into As (V); and subsequently enhanced better adsorption of the arsenic.

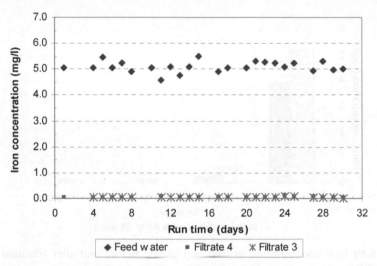

**Figure 6.21** Iron concentration in feed and filtrate of filter 4 having a polishing sand layer of bed depth 10 cm and size fraction 0.5 mm to 1.0 mm.

# 6.6    Conclusions

- The UNESCO-IHE family filter filled with IOCS treating model water with high ammonium, iron, manganese and arsenic concentrations, removed manganese adequately during the first 10 days of continuous operation. Prolonged filter run resulted in manganese breakthrough and subsequently adsorbed manganese was released from the IOCS. These observed phenomena are attributed to the oxidation of ammonium by Nitroso bacteria, resulting in an anoxic condition in the filter. Under that condition manganese was not removed adequately; on the longer term, accumulated manganese was released, likely due to reduction of manganese dioxide to manganese (II).

- The Family filter having a polishing layer of a manganese mineral (Aquamandix) on top, removed consistently manganese below the 0.4 mg/l WHO guideline. Adequate removal of manganese is attributed to the high adsorption capacity for manganese (II) of the polishing layer.

- Post sand filtration after aeration did not remove manganese from the effluent of the Family filter within seven weeks operation at pH 6.8. Manganese removal started after this period indicating that the formation of catalytic manganese oxides at about pH 7 is very slow. This is in concert with the results in Chapter 5 and findings in practice.

- A Family filter with a layer of fine sand on the top of IOCS fed with model water without ammonium, showed significant higher manganese removal efficiency. This result demonstrates that the presence of ammonium reduced the removal of manganese and caused the release of manganese by creating anoxic conditions, due to bacterial oxidation of ammonium.

- High ammonium content (4 mg/l) of the model water caused fast depletion of dissolved oxygen within both filters (i.e. Family filters 1 and 2) and the establishment of anoxic conditions with the attendant high nitrite concentrations in filtrates exceeding 0.2 mg/l (WHO guideline). The formation of nitrite and the inability of the filters to convert it to nitrate is attributed to the absence of sufficient oxygen in the feed water. A polishing filter fed with aerated effluent removed the nitrite adequately.

- Arsenic removal in the Family filter was initially adequate; however soon its performance reduced. Frequently, the arsenic concentration in the effluent exceeded the level of 10 µg/l (WHO guideline). The observed deterioration of the effluent quality is attributed to the escape of ferric hydroxide flocs. These flocs contained relatively high amounts of arsenic. This dominant cause of deterioration of the effluent quality is clearly illustrated by the facts that:

  - a polishing layer of sand placed on top of the Family filter removed the arsenic to levels below 10 µg/l;
  - post-treatment of aerated effluent of the Family filter with a polishing sand filter had the same effect of arsenic removal;
  - filtration of the effluent of the Family filter through a membrane filter (0.45 µm) reduced the arsenic concentration substantially.

- The Family filter equipped with a polishing layer of Aquamandix on top, removed arsenic consistently below 10µg/l throughout the experimental period. These results are due to the effective removal of ferric hydroxide flocs and adsorption of arsenic by the Aquamandix polishing layer.

- Iron removal in the Family filter equipped with IOCS only was initially adequate, however after several weeks the concentration exceeded 0.3 mg/l. Treating the aerated effluent with a filter filled with fine sand reduced the iron concentration adequately. The same effect was achieved by placing a polishing layer of fine sand on top of the IOCS in the Family filter.

- The Family filter with addition of Aquamandix layer on the top of IOCS showed adequate removal of arsenic, iron, and manganese when treating model

water with high $NH_4^+$. A polishing, post-sand filter following an aeration step
will be required to remove nitrite.

- A polishing filter filled with fine grains of Aquamandix, treating the aerated
  effluent of a Family filter with IOCS only, is recommended to ensure adequate
  removal of manganese and nitrite. In addition this post treatment will improve
  the removal of arsenic and iron.

# 6.7 References

Banerjee, K. and Amy G.L. 2004 Adsorption of Arsenic Onto Granular Ferric Hydroxide in
    Presence of Silica and Vanadium. Proc. 2004 AWWA WQTC Conf., San Antonio.
Bajpai, S and Chaudhuri, M. 1999 Removal of arsenic from groundwater by manganese dioxide
    coated sand. J. of Env. Eng., 125(8), 782 – 784.
Barua, R. 2006; Long term monitoring and optimisation of manganese removal with IHE Family
    Filter. M.Sc. Thesis; Department of Urban Water Sanitation, - UNESCO-IHE, Institute
    for water education. pp 108.
Bissen, M. and Frimmel, F.H. 2003b Arsenic – a review. Part II: Oxidation of arsenic and its
    removal in water treatment, Acta hydrochim. Hydrobiol., 31(2), 97 – 107.
Borho, M. and Wilderer, P. 1996 Optimized removal of arsenate (III) by adaptation of oxidation
    and precipitation processes to the filtration step. Wat. Sci. Tech., 39(9): 25 – 31.
Bock, E., Schmidt, I., Stuven, R. and Zart, D. 1995 Nitrogen loss caused by Denitrifying
    Nitrosomonas cells using ammonium or hydrogen as electron donors and nitrite as
    electron acceptor. Archive Microbiology vol. 163, pp. 16 – 20.
Driehaus, W., Seith, R., and Jekel, M. 1995 Oxidation of As(III) with manganese oxides in water
    treatment, Wat. Res., 29(1): 297 – 305.
Gable, J., and Fox, P. 2003 Sustainable Nitrogen removal by anaerobic ammonia oxidation.
    Proceedings of the 76th annual Water Environment Federation Conference, Los Angeles,
    CA.
Hem, J. D. 1977 Reactions of metal ions at surfaces of hydrous iron oxide.Geochim.Cosmochim.
    Acta, 41, pp. 527 – 538, Oxford, New Yourk. Paris, Frankfurt: pergamon..
Hug, S.J. and Leupin, O. 2003 Iron-catalyzed oxidation of arsenic(III) by oxygen and by hydrogen
    peroxide: pH-dependent formation of oxidants in the Fenton reaction. Environ. Sci.
    Technol.,37, 2734 – 2742.
Jetten, M.S.M., Strous, M., van de Pas-Schooen, K.T., Schalk, J., van Dongen, U.G.J.M., van de
    Graaf, A.A. Logemann, S., Muyzer, G., van Loosdrecht M.C.M. and Kuenen 1999. The
    anaerobic oxidation of ammonium. FEMS Microbiology Reviews, vol. 22, pp 421 – 437.
Lee, Y.; Um, I.H. and Yoon, J., 2003 Arsenic (III) Oxidation by Iron(VI) (ferrate) and subsequent
    removal of arsenic(v) by Iron(III) coagulation. Envir. Sci. & Technol., 37: (24): 5750.
Pal, B.N. 2001 Granular ferric hydroxide for elimination of arsenic from drinking water.
    www.unu.edu/env/Arsenic/Pal.pdf
Painter, H. A. 1970 A review of literature on inorganic nitrogen metabolism I microorganisms.
    Water Research, vol. 4, p 393.
Petrusevski, B., Boere, J., Shahidullah, S.M., Sharma, S.K. and Schippers, J.C. 2002 Adsorbent
    based point-of-use system for arsenic removal in rural areas. J. Water Supply Res.
    Technol. 51, 135 – 144.
Petrusevski, B.; Sharma, S.K.; Krius, F.; Omeruglu, P. and Schippers, J.C. 2003 Family filter with
    iron-coated sand: solution for arsenic removal in rural areas. J. of Wat. Sc. and Technol.
    Water Supply. 2,(5 – 6) 127 – 133.

Petrusevski, B., Sharma, S.K., van der Meer, W., Kruis, F., Khan, M., Barua, M. and Schippers,
    J.C. 2008 Four Years of Development and Field-testing of IHE Arsenic Removal Family
    Filter in Rural Bangladesh. *Water Science and Technology*, 58 (1) 53-58.
Salehuddin, A. K. M. (2005) Long-Term Performance Monitoring and Optimisation of IHE
    Family Filter for Arsenic Removal in Bangladesh, M.Sc. Thesis (SE 05-021), Delft, the
    Netherlands.
Sharma, S.K. 2002 Adsorptive iron removal from groundwater. PhD Thesis. UNESCO-IHE,
    Instistute for Water Education and Wageningen University. Swets and Zeitlinger B.V.,
    Lisse.
Stumm, W. & Morgan, J.J. (1996) Aquatic Chemistry: Chemical equilibria and rates in natural
    waters, $3^{rd}$ edition. Wiley, New York, USA, pp 462.
Tchobanoglous, G., Burton, F.L. and Stensel, H. D. 2003 Wastewater Engineering:Treatment and
    Reuse. $4^{th}$ edition. Metcalf & Eddy Inc. Tata McGraw – Hill Publishing Co. Limited.
    New Delhi, New York, Milan, Boston. pp. 612 – 613.
Thirunavukkarasu, O.S.; Viraraghavan, T and Subramanian, K. S. 2003 Arsenic removal from
    drinking water using granular ferric hydroxide. *Water SA*, 29:2:161.
Teske, A., Alm, E., Regan, Toze, S., Rittman, B.E. and Stahl, D.A. 1994 Evolutionary
    relationships among ammonia and nitrite-oxidizing bacteria. *Journal Bacteriology*, vol.
    176, pp 6623 – 6630.
Wagner, M., Rath, G. Amman, R., Koops, H.P. and Schleifer, K.H. 1995 In situ identification of
    ammonia-oxidizing bacteria. *System Applied Microbiology*, vol. 18, pp. 251-264.

# APPENDIX 6.1

**Table 2** Calculations on ammonium removal, nitrite and nitrate formation and oxygen consumption.

|  | Filter 1 | Filter 2 | Filter 3 |
|---|---|---|---|
| $NH_4^+$ in (mg/l) | 3.8 | 3.8 | 0.8 |
| $NH_4^+$ out (mg/l) | 0.8 | 0.7 | 0.0 |
| $NH_4^+$ removed (mg/l) | 2.9 | 3.0 | 0.8 |
| $O_2$ available in feed water (mg/l) | 9.5 | 9.5 | 6.6 |
| *$O_2$ (mg/L) require to convert 1 mg$NH_4^+ \to NO_2^-$ | 2.7 | 2.7 | 2.7 |
| $O_2$ (mg/l) required for the convertion of the removed $NH_4^+ \to NO_2^-$ | 10.1 | 10.4 | 2.8 |
| By calculation, the amount (mg) of **$NO_2^-$ that can be produced from the removed $NH_4^+$ | 7.5 | 7.7 | 2.1 |
| $NO_2^-$ measured in filtrates of the experiment (mg/l) | 0.4 | 2.2 | 0 |
| $NO_2^-$ removed (mg/l) | 7.1 | 5.6 | 2.1 |
| *$O_2$ (mg/l) required for the convertion of 1 mg $NO_2^- \to NO_3^-$ | 0.4 | 0.4 | 0.4 |
| $O_2$ mg/l required for convertion of removed $NO_2^- \to NO_3^-$ | 2.5 | 2.0 | 0.7 |
| Total oxygen required for the oxidation of $NH_4^+$ to $NO_3^-$ (mg/l) | 12.5 | 12.3 | 3.6 |
| Oxygen Required – Oxygen available (mg/l) | 3.0 | 2.8 | -3.0 |
| Possibility of total oxidation of ammonium to nitrate | No | No | Yes |

# CHAPTER SEVEN

# OXIDATION OF ADSORBED FERROUS AND MANGANESE: KINETICS AND INFLUENCE OF PROCESS CONDITIONS

Part of this chapter was published as:
Buamah, R. Petrusevski, B. and Schippers, J.C. (2008) Oxidation of adsorbed ferrous and manganese: kinetics and influence of process conditions. In: Proceedings of *the UCOWR / NIWR Conference at North Carolina, Durham (U.S.A)* 21 – 23 July, 2008.

Another part has been accepted for publication as
Buamah, R. Petrusevski, B. and Schippers, J.C. (2009) The influence of process conditions on the oxidation kinetics of adsorbed ferrous. *Journal of Water Science and Technology: Water Supply* (in press).

# Abstract

For the removal of iron and manganese from groundwater, aeration followed with rapid (sand) filtration is frequently applied. Iron removal in this process is achieved through oxidation of $Fe^{2+}$ in aqueous solution followed by floc formation as well as adsorption of $Fe^{2+}$ on the surface of filter media followed by oxidation. Manganese removal is mainly achieved through $Mn^{2+}$ adsorption on filter media followed by its oxidation.. The rate of oxidation of the adsorbed $Fe^{2+}$ and / or $Mn^{2+}$ on the filter media plays an important role in this removal processes. For this reason, preliminary experiments have been conducted to investigate the effect of pH on the rate of oxidation of adsorbed $Fe^{2+}$ and $Mn^{2+}$. For this purpose $Fe^{2+}$ has been adsorbed, under anoxic conditions, on iron oxide coated sand (IOCS) in a short filter column and subsequently oxidized by feeding the column with aerated water. IOCS has been formed in rapid sand filters removing iron from ground water in a full scale plant. $Mn^{2+}$ has been adsorbed on Aquamandix ( a synthetic and predominantly manganese dioxide filter medium). These media have been chosen since in previous related experiments IOCS demonstrated the highest adsorption capacity for $Fe^{2+}$ and Aquamandix for $Mn^{2+}$.

Ferrous ions adsorbed on IOCS in the short column at pH 5, 6, 7 and 8 demonstrated consumption of oxygen, when aerated water was fed into the column. The oxygen uptake was at pH 7 and 8 faster than at pH 5 and 6. However the difference was much less pronounced than expected.

The less pronounced difference is attributed to the pH buffering effect of the IOCS. Using aerated feed water at pH values 5, 6 and 7, the pH in the effluents were higher than in the influents, while pH drop should occur, because of the oxidation of adsorbed $Fe^{2+}$. The unexpected increase in pH is attributed to the presence of calcium and /or ferrous carbonate in IOCS. At pH 8, the effluent pH dropped as expected.

Adsorbed $Mn^{2+}$ on Aquamandix did not demonstrate any uptake of oxygen within the test period of 120 minutes. This unexpected result indicates that $Mn^{2+}$ being adsorbed onto the Aquamandix will not be oxidized rapidly. This finding makes this medium attractive as an alternate filter medium only if an adsorptive catalytic manganese oxide coating will be formed rapidly during operation.

**Keywords**: Oxidation, ferrous, manganese, adsorption, kinetics, IOCS, Aquamandix

# 7.1    Introduction

Ground waters containing $Fe^{2+}$ and $Mn^{2+}$ are frequently treated by aeration followed by one or two stage rapid (sand) filtration. In the iron removal from groundwater two chemical / physical mechanisms are involved namely:

- oxidation of $Fe^{2+}$ to $Fe^{3+}$ which hydrolyses immediately to $Fe(OH)_3$ and forms flocs (flocculation removal mode);

- adsorption of $Fe^{2+}$ on the surface of the filter media, followed by oxidation to $Fe^{3+}$ (still adsorbed), hydrolyzing and formation of a dense coating. This coating has a higher adsorption capacity than the filter media (adsorptive removal mode).

The pH, oxygen concentration and pre-oxidation time determine to what extend the different modes occur. Studies of Sharma (2002) showed that development of head loss is minimal when adsorptive iron removal occurs and consequently the required frequency of backwashing can be reduced substantially.

The physical / chemical removal mechanism of $Mn^{2+}$ is expected to be adsorption of $Mn^{2+}$ on the surface of the media followed by oxidation of the adsorbed $Mn^{2+}$ to $Mn_3O_4$ and $MnO_2$. The adsorbed and subsequently oxidized $Mn^{2+}$ acts as newly created adsorbent with high adsorption capacity. The adsorbed and oxidized manganese form a dense coating. Manganese is consequently removed through Adsorptive removal mode.

Aside using aeration to oxidize the dissolved $Fe^{2+}$ and $Mn^{2+}$, chemical oxidants like $KMnO_4$, $Cl_2$, $ClO_2$, $NaOCl$, and $O_3$ are used in practice as well. Beside sand, several other filter media including sand coated with iron oxides and / or manganese, zeolites of volcanic origin, manganese dioxide or manganese greensand are used in rapid filtration. Two different modes of (chemical) oxidation can be identified namely:

- Continuous aeration and / or continuous addition of an oxidant
- No aeration and intermittent oxidation with an oxidant.

The oxidation of the adsorbed $Fe^{2+}$ and $Mn^{2+}$ is essential in the adsorptive mode of removal of these compounds. The knowledge on the rate of oxidation and in particular the effect of pH is, however, very limited. This study therefore focuses on the rate of oxidation of adsorbed $Fe^{2+}$ and $Mn^{2+}$. For this purpose preliminary experiments were conducted to investigate the effect of pH on the rate of oxygen uptake in short column tests with iron oxide coated sand (IOCS) loaded with Fe (II) and Aquamandix (a crushed manganese dioxide medium) loaded with Mn (II). IOCS was used for the adsorption / oxidation of Fe (II) since this medium had the highest adsorption capacity for Fe (II) (Sharma, 2002) while Aquamandix has shown the highest adsorption capacity for Mn (II) (Chapter 4 of this thesis).

## 7.2    Theoretical background for kinetics:

The oxidation kinetics of dissolved $Fe^{2+}$ in aqueous solution is depicted by the following equation:

$$4 Fe^{2+} + O_{2\,(g)} + 10 H_2O \rightarrow 4 Fe(OH)_{3(s)} + 8 H^+ \tag{7.1}$$

At a $pH \geq 5$, the following rate law is applicable (Stumm and Lee, 1961):

$$\frac{d\ [Fe(II)]}{dt} = -\ k\ pO_2[OH^-]^2[Fe(II)] \tag{7.2}$$

where:
k = rate constant $[l^2/(mol^2.atm.min)]$
$pO_2$ = partial pressure of oxygen (atm)
$[OH^-]$ = hydroxide ion concentration, mol /l
$[Fe^{2+}]$ = $Fe^{2+}$ concentration, mol /l

The effect of pH on the rate of oxidation is very pronounced e.g. a change in pH of one unit, results in a change in oxidation rate with a factor 100. Factors that influence the rate of oxidation of $Fe^{2+}$ aside the pH and oxygen concentration include ferric ions, alkalinity, temperature, organic matter, silica, copper, manganese and cobalt (Stumm and Lee, 1961; Ghosh et al. 1996). According to Tufekci and Sarikaya (1996), Fe (III) plays a significant role in the oxidation process as well, and it is depicted in the equation below.

$$-\ \frac{d\ [Fe(II)]}{dt} = (k^\theta + k^1[Fe\,(III)])[Fe\,(II)] \tag{7.3}$$

In the presence of Fe(III) the oxidation of Fe(II) proceeds along two pathways:

• the homogenous reaction pathway that takes place in solution;
• the heterogeneous reaction that occurs on the surface of ferric hydroxide precipitates; indicating that this part of the oxidation process is autocatalytic.

The rate constant $k^\theta$ for homogenous reaction is equal to $K_o[O_2][OH^-]^2$ and $k^1$ for the heterogeneous reaction is determined by $k_{so}[O_2]K_e\ /\ [H^+]$; $K_o$ and $k_{so}$ being real rate constants and $K_e$ being the equilibrium constant for the adsorption of iron (II) onto iron (III) hydroxides. Since $k^1$ is proportional to $K_e$, the rate of oxidation of adsorbed Fe (II) most likely depends on the pH, because the adsorption of Fe (II) on media depends strongly on the pH (Sharma, 2002).

To explain the mechanisms involved in the adsorptive removal mode of iron, the following illustrative equations have been proposed (Barry et al, 1994; Sharma, 2002):

$$\equiv S\text{-}OH + Fe^{2+} \leftrightarrow \equiv S\text{-}OFe^+ + H^+ \tag{7.4}$$

Where ' $\equiv S$ 'represents the surface of the filter media. In the presence of oxygen and at the appropriate pH, normally > pH 5, the adsorbed $Fe^{2+}$ is oxidized and hydrated thereby creating new adsorptive surface as follows:

$$\equiv S\text{-}OH + Fe^{2+} + \tfrac{1}{4}\,O_2 + {}^3/_2\,H_2O \rightarrow\ \equiv S\text{-}OFe \begin{subarray}{l} OH \\ \\ OH \end{subarray} + 2\,H^+ \tag{7.5}$$

The removal of dissolved $Mn^{2+}$ from groundwater is generally accomplished by adsorption on filter media at appropriate pH (i.e. > pH 7) followed by oxidation with oxygen present in the water or with chemicals e.g. potassium permanganate. The following manganese oxidation equation has been proposed by Stumm and Morgan (1996):

$$\frac{d[Mn(II)]}{dt} = - k_o\,[Mn(II)] + k_1[Mn(II)][MnO_2] \qquad (7.6)$$

where:
$k_o = k\,PO_2\,.\,[OH^-]^2$
$k$ = reaction rate constant $[l^2/(mol^2.atm.min)]$
$k_1$ = reaction rate constant $[l^3/(mol^3.atm.min)]$
$PO_2$ = Partial pressure of oxygen (atm)

The rate of adsorbed Mn (II) oxidation does depend upon the oxygen concentration and pH as well as the Mn (II) and $MnO_2$ concentrations. This implies that the reaction is autocatalytic. The effect of pH on the oxidation in aqueous solution is as pronounced as in the oxidation of Fe (II). Assuming that the mechanism of Mn (II) oxidation is similar to equation 7.3, $k_1$ will depend on pH as well, since $k_1$ will be proportional to adsorption equilibrium constant for manganese, which usually depend strongly on pH (Chapter 4).

At pH values below 9 the rate of Mn(II) oxidation in aqueous solutions is very low (Stumm & Morgan, 1996). However, in rapid (sand) filters manganese removal has been observed down to pH 6.9. This phenomenon indicates that the rate of oxidation of adsorbed Mn (II) is much higher than Mn (II) in aqueous solutions. So the process is catalytic / adsorptive and attributed to manganese oxides e.g. $Mn_3O_4$ and $MnO_2$.

# 7.3    Material and Methods

The objective of the experiments was to reveal the rate of oxidation of adsorbed $Fe^{2+}$ and $Mn^{2+}$ at different pH levels. For the oxidation of adsorbed $Fe^{2+}$, iron oxide coated sand (IOCS) was placed in a short filter column and loaded with $Fe^{2+}$ under anoxic conditions. Subsequently the adsorbed $Fe^{2+}$ was exposed to oxygen containing water and the oxygen consumption in the short filter column monitored by periodically determining the dissolved oxygen content of the influent and the effluent of the column.

In case of the oxidation of adsorbed $Mn^{2+}$, Aquamandix (particle size: 1.0 – 1.2 mm) was used as the filter medium.

## 7.3.1 Preliminary tests

### 7.3.1.1 Characterization of IOCS

Iron oxide coated sand (IOCS) from the Noord Bargeres groundwater treatment plant in the Netherlands was used as filter media in this study. The chemical composition of the

IOCS was determined by digesting with analytical grade concentrated $HNO_3$ (65%) and $HClO_4$ in accordance with the protocol in the Standard Methods (AWWA) (2005). After sieving, IOCS grains of particle size 2 – 4 mm were used for the pilot column experiments. The bulk density of the IOCS grains was determined from its mass and volume.

### 7.3.1.2 Determination of the pore volume of the mineral coating of the IOCS

250 ml of water was added to 100g of the IOCS (dried at 110°C) in a measuring cylinder and closed tightly to prevent evaporation. The level of the water in the cylinder was noted immediately after addition of the water and daily during one week. The decrease in the level of water in the cylinder recorded gave the pore volume of the IOCS for the given mass.

## 7.3.2 Experimental set up

### 7.3.2.1 De-aeration column

The pilot set-up comprises a de-aeration column connected in series with a short column and oxygen meters (Figures 7.1and 7. 2). A 4m transparent perspex column (i.e. the

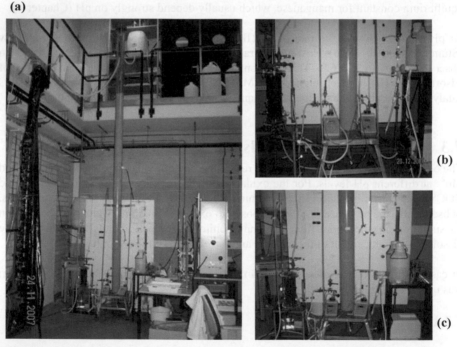

**Figure 7.1** The Photographs of the column set: (A). A full view of set-up (B) A closer view with short column filled with Aquamandix (C) short column filled with IOCS.

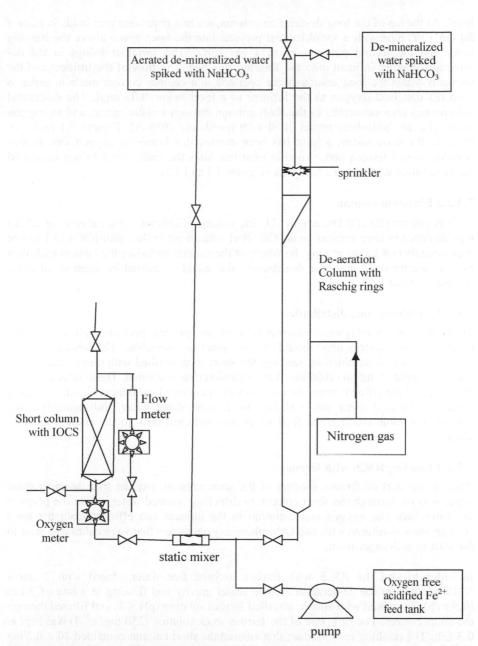

**Figure 7.2** Schematic diagram of pilot column set-up.

de-aeration column) with internal diameter of 0.13 m, was filled to about 70% of its height with 3 mm ceramic Raschig rings. At about 1 m height of the long de-aeration column, is a port used for infusing nitrogen gas into the column, just above the water level. At the top of the long de-aeration column, are two ports; one port holds in place a 50 cm PVC pipe with a sprinkler that projects into the open space above the Raschig rings and the second port serves as a gas valve. The sprinkler brings in the de-mineralized water influent onto the Raschig ring bed. The flow of the influent and the nitrogen within the long column were operated in a counter current mode in order to strip off dissolved oxygen in the influent to a level below 0.05 mg/l. The de-aerated column has been connected to the short column through a static mixer, and an oxygen meter (i.e. an Orbisphere model 3650 with membrane 2956 A) (Figures 7.1 and 7.2). Prior to the static mixer, a joint has been created that brings in oxygen free ferrous solution from a ferrous tank. Another joint that links the static mixer brings in aerated de-mineralized water from a feed tank (Figures 7.1 and 7.2).

### 7.3.2.2 Filtration column

A short column (ID of 0.1m, length of 0.3m, volume of column + conical ends → 2.57L) was designed to have conical ends. The short column was filled with IOCS (3.1 kg) or Aquamandix (4.8 kg) to occupy the whole of the column including the conical ends thus minimizing the residence time distribution, that could be caused by supernatant water (Figures 7.1 and 7.2).

### 7.3.2.3 Residence time distribution

To enable an unambiguous interpretation of the oxygen breakthrough curves, the residence time distribution should be as small as possible. The residence time distribution was determined by feeding, the short column filled with clean sand, with a sodium chloride solution (1000 mg/l) at a filtration rate of 0.9 m/h. The conductivities of the feed and the effluent from the column were monitored continuously. After a stable reading , the feed water was switched to de-mineralized water. Subsequently, after obtaining a stable reading, the feed water was switched back to the sodium chloride solution.

### 7.3.2.4 Loading IOCS with ferrous

Prior to the start of ferrous feeding of the short column, oxygen free de-mineralized water was run through the short column to dispel any aerated water within the pores of the filter bed. The oxygen concentration in the influent and effluent from the short column were monitored with two Orbisphere oxygen meters that were calibrated prior to the start of each experiment.

In order to load the IOCS with ferrous, oxygen free water (dosed with 2 mmol $NaHCO_3$ /l) from the de-aeration column under gravity and flowing at a rate of 7 L/h (0.9 m/h) was spiked with anoxic acidified ferrous solution (pH < 2) and filtered through the short column. The flow rate of the ferrous stock solution (250 mgFe$^{2+}$/l) was kept at 0.3 L/h. The resulting feed mixture that entered the short column contained 10 ± 0.2 mg Fe$^{2+}$/l and had a pH of 7.0 ± 0.2. The IOCS in the short filtration column was loaded with ferrous (10 mg/l) under anoxic conditions for 8½ hours.

### 7.3.2.5 Oxidation of adsorbed ferrous

After the ferrous loading the short column was fed with aerated water having dissolved oxygen concentration (DO) of $8.3 - 8.6$ mg $O_2/l$. The aerated model water was prepared by dosing de-mineralized water with 2 mmol $NaHCO_3$ /l and aerating the mixture for about one hour. The aerated water was allowed to stay for about 12 hours to equilibrate prior to application. The pH of the aerated water was adjusted using 6N HCl / 3N NaOH to a particular pH (5, 6, 7 or 8). The anoxic ferrous loaded short column was fed with the aerated water and concurrently the dissolved oxygen and pH of the influent and effluent were monitored periodically with the pre-calibrated oxygen meters and pH meters. Periodic samples were taken from the short column effluent for analysis to monitor the total iron and calcium content. The flow rate in the short column was maintained at 7 L/h (corresponding to filtration rate of $V_f = 0.9$ m/h).

Using a flow rate of 7 L/h ($V_f = 0.9$ m/h), the water replacement or contact time in the short column was calculated and found to be 8.8 minutes. The volume of the voids within the filter bed was determined to be 1.03 litres; meaning that at the start of the ferrous oxidation process, this volume (i.e. 1.03 litres) of aerated water is required to replace the existing anoxic water.

### 7.3.2.6 Loading and oxidation of manganese (II) adsorbed onto Aquamandix

Aquamandix is predominantly a crushed manganese dioxide material with some iron, silica and alumina (Table 7.1). Aquamandix has been shown to have a higher manganese adsorption capacity than IOCS (refer chapter 4). Consequently adsorbed manganese (II) oxidation experiments were conducted using Aquamandix (particle size: $1.0 - 1.2$ mm) in place of IOCS. Similar protocol as described above for the ferrous loading was applied for the manganese loading with the following few changes. The anoxic $Mn^{2+}$ fed into the short column with Aquamandix was 4 mg$Mn^{2+}$/l, and the pH was 7. The hydrogen carbonate concentration of the anoxic feed was 2 mmol $NaHCO_3/l$. The concentration of 4 mg$Mn^{2+}$ /l was chosen to avoid manganese carbonate precipitation (chapter 3). After loading the column with $Mn^{2+}$, oxidation of the adsorbed $Mn^{2+}$ was carried out with aerated de-mineralized water at pH 6 and 8.

**Table 7.1** The chemical composition of Aquamandix.

| Inorganic ion | % chemical composition of Aquamandix(mg/g media) |
|---|---|
| $Fe_2O_3$ | 6.2 |
| $MnO_2$ | 78.0 |
| $SiO_2$ | 5.2 |
| $Al_2O_3$ | 3.1 |

(source: Aqua Techniek – 2007)

**7.3.2.7 Oxygen meter calibration and blank test with IOCS**

Oxygen concentrations were monitored with Orbisphere oxygen meters (i.e. Orbisphere model 3650 with membrane 2956 A) which were calibrated before and after each experiment. The detection level of these meters is 5 µg/l. Experiments were conducted to determine the response time of the oxygen meters. The response time of the oxygen meters turned out to be 50 seconds under both oxic and anoxic conditions. To get an impression of 'tailing' during the oxidation experiments, before each test, a blank experiment has been conducted. For this purpose, anoxic feed water has been passed through the short column for eight and half hours and subsequently replaced with aerated water.

# 7.4    Results and discussions

## 7.4.1 Characterization of the IOCS media

The preliminary characterization test gave the results shown in Table 7.2. The bulk density of the IOCS was found to be 1.3 g/cm$^3$ (i.e. 1300 Kg/m$^3$). For a unit mass of the IOCS, 17% by mass of it was found to be constituted by the mineral coating. The pore volume analysis showed that, for a given volume of the IOCS, 18.2 % comprise the internal pores within the mineral coating of the media.

From the high porosity of the mineral coating, it can be deduced that rate of oxidation of adsorbed iron could be influenced by the rate of diffusion of oxygen and other ions / molecules influencing the pH like H$^+$, HCO$_3^-$, and CO$_2$ since Fe$^{2+}$ is assumed to be adsorbed on the external surface and within the pores of the mineral coating. IOCS has been reported to have a high Fe$^{2+}$ and Mn$^{2+}$ adsorption capacity (Sharma, 2002; refer Chapter 3). The high adsorption capacity for Fe$^{2+}$ and Mn$^{2+}$ is attributed to the high iron and manganese content of the IOCS respectively (Table 7.2).

Table 7.2 Chemical composition of IOCS.

| Inorganic ion | IOCS composition (mg/g media) |
|:---:|:---:|
| Fe | 367.60 |
| Mn | 19.30 |
| Ca | 9.45 |

## 7.4.2 Residence time distribution in filtration column

The experiments carried out with a column filled with virgin quartz sand (particle size: 0.6 – 1.2 mm) gave rise to an asymmetrical plot with a long tailing indicating a longer residence time distribution (Figure 7.3). This may be due to the relatively more angular conical ends (about 45°) that were used initially for the filter column. Subsequently the conical ends were replaced with less angular conical ends (about 20°) and filled with the filter media to reduce the residence time distribution to a minimum.

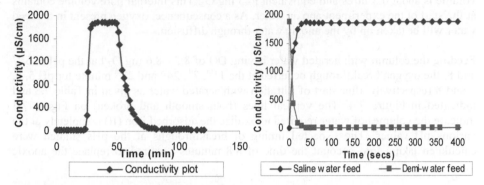

**Figure 7.3** Residence time distribution curve.

## 7.4.3 Column experiments – adsorbed ferrous oxidation

### 7.4.3.1 Oxygen breakthrough curves

As expected almost all $Fe^{2+}$ in the feed water was adsorbed during the loading at pH 7, since the theoretical adsorption capacity of the column was about 3000 mg (capacity calculated using equilibrium ferrous concentration of 10 mg/l). Amounts of 558 - 644 mg of ferrous were adsorbed during the loading phase preceeding the oxidation experiments at the various pHs. The Table 7.3 gives a mass balance summary of the amounts of the adsorbed ferrous and their respective amounts of oxygen consumed during the oxidation experiments at breakthrough (a detailed mass balance table is given in the appendix 7.1). Breakthrough point is considered to have been reached when there is a decline in the oxygen consumed in the filter. In figures 7.4 a, b and c, the oxygen breakthrough curves are given as a function of the filtered volume, for the blank and the adsorption / oxidation tests. The adsorption / oxidation curves are showing a remarkable tailing. This phenomenon is attributed to the limitation in rate of oxidation and mass transfer (diffusion) outside and inside the IOCS grains, and short circuiting in the column.

**Table 7.3** Summarized mass balance of adsorbed ferrous oxidation.

| pH | Net Fe(II) Adsorbed (mg) | Stoichiometric amount of $O_2$ required for oxidation (mg$O_2$) | Actual amount of oxygen consumed (mg$O_2$) at breakthrough | Experimentally determined breakthrough time (min) | % adsorbed Fe (II) oxidized at breakthrough |
|---|---|---|---|---|---|
| 5 | 602.2 | 84.3 | 10.8 | 11.0 | 12.8 |
| 6 | 582.0 | 81.5 | 7.0 | 7.0 | 8.6 |
| 7 | 557.7 | 78.1 | 24.0 | 24.0 | 30.7 |
| 8 | 644.0 | 90.2 | 26.3 | 27.0 | 29.1 |
| 7 higher $V_f$ | 595.2 | 83.3 | 3.1 | 1.5 | 3.7 |

The blank test gives an indication of the tailing due to the limitation in mass transfer and short circuiting under conditions when no oxygen is consumed. The internal pore volume is about 0.3 litres and equivalent to 3 mg $O_2$. This internal pore volume contains at the start of the experiment anoxic water. As a consequence, oxygen present in the feed water will be taken up by the anoxic water through diffusion.

Feeding the column with aerated water having DO of 8.3 - 8.6 mg $O_2$/l at the pH 5, 6, 7 and 8, the oxygen breakthrough occurred at the 11th, 7th, 24th and 27th minute for pH 5, 6, 7 and 8 respectively after start of filtration with aerated water as given in Table 7.3 and indicated in Figure 7.4. The vertical lines (both smooth and broken) on Figure 7.4 indicate the volumes of water required to oxidize the adsorbed iron (II) completely at the various pH values. The expected timing of breakthroughs at the pHs studied were calculated taking into account the time of 8.8 minutes required to replace the anoxic

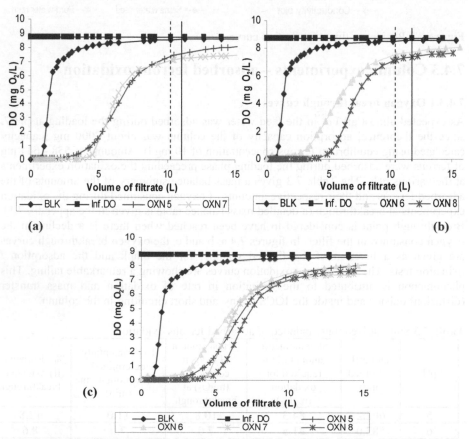

**Figure 7.4** Oxygen breakthrough curves for blank test and oxidation of adsorbed Fe (II). (a) Oxidation at pH 5 and 7 with the expected breakthrough lines: smooth line →pH 5 and broken line→ pH 7. (b) Oxidation at pH 6 and 8 with the breakthrough lines: broken line → pH 6 and smooth line → pH 8. (c) The combined plots.

water in the IOCS pores. From the amounts of oxygen consumed, it was found that at breakthrough 12.8%, 8.6%, 30.7 and 29.1% of the adsorbed ferrous were oxidized at pH 5, 6, 7 and 8 respectively. The filtered volumes at breakthrough for pH 5 and 6 were much smaller than at pH 7 and 8 (Figure 7.4c). In addition from figure 7.4, it is also observed that the slope of the pH 7 and 8 oxidation curves at the breakthrough is steeper than that of the pH 5 and 6 suggesting a higher rate of oxidation of the adsorbed ferrous at the elevated pH.

Despite the increase in rate of oxygen uptake as pH increases from 6 to 8, the increment does fall short of expectation, when compared with ferrous oxidation in a homogenous aqueous system. In an aqueous solution an increment of the factor 10 000 in the rate of oxidation can be derived from equation 7.3 when pH increases from 6 to 8. However there is no information available on the effect of pH on the rate of oxidation of adsorbed Fe(II).

In addition a number of other factors might have an impact on the rate of oxygen uptake, which might explain the limited difference between observed $Fe^{2+}$ oxidation kinetics at pH 6 and 8:

- External mass transfer of oxygen, $OH^-$ and $HCO_3^-$ ions from the bulk of the water to the surface of the IOCS grains could influence the oxidation through pH changes. $OH^-$ and $HCO_3^-$ will be transferred to the grains surface since the pH would drop inside the pores due to the oxidation of adsorbed Fe (II); $H^+$ ions will go in the other direction.
- Internal pore diffusion of oxygen, $H^+$, $HCO_3^-$ and $OH^-$ ions may influence oxygen uptake.
- Mass transfer at the surface of the pores, where reduction in pH occurs, due to oxidation of adsorbed Fe (II). Mass transfer at this position might be a limiting factor as well (Sharma 2002).

### 7.4.3.2 Effect of filtration rate

To investigate the effect of filtration rate on the oxidation of adsorbed $Fe^{2+}$, an experiment was done using a higher filtration rate of 2 m/h (corresponding to a flow rate of 14.2 l/h) and aerated water with DO of 8.34 $mgO_2/l$ in anticipation that more oxygen will be taken up in a same period of time. The IOCS was pre-loaded with a similar amount of $Fe^{2+}$. Using the aerated water DO value, the filtration rates and figure 7.5, it can be derived that the oxygen uptake within a period of (say) 34 minutes in which 4 and 8 litres (net) passed the filter operated at filtration rate of 0.9 m/h and 2.0 m/h respectively, is somewhat higher at the higher filteration rate. This suggests that the mass transfer depends on the rate of filtration. However during an important period in the test at low filtration rate the water is anoxic which means that more oxygen could have been consumed than available in the water. As a consequence it can not be concluded that external mass transfer at pH 7 played a role as a limiting factor in oxygen up take.

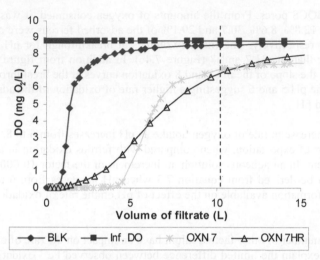

**Figure 7.5** Effect of higher rate of filtration breakthrough curve. BLK - Blank at pH 7; OXN 7HR - high filtration rate oxidation at pH 7; OXN 7 - oxidation at pH 7 and Inf. DO. - Influent Dissolved Oxygen content at pH 7.

### 7.4.3.3 Impact of IOCS on pH

In verifying the pH of the effluents of the column during the different experiments, it was observed that the pH levels deviated substantially from the influent pH. It was expected that the pH would drop because of the oxidation of adsorbed Fe (II). However during the ferrous oxidation phase at pH 5 and 6, the pH of the effluent from the short column did not remain constant but increased remarkably up to 7.2 (Figure 7.6). The observed increased level of the pH in the effluent of the column indicates that the pH in the column was certainly above the pH of the influent. As a consequence it is expected that the rate of oxidation of adsorbed Fe (II) has been increased as well. Since the pH of the effluent of the column had the similar level in both tests, the difference in oxidation rate was minimized.

On the contrary, in the test conducted at pH 8, the pH of the effluent dropped to 7 – 7.5 (Figure 7.6). In this case the rate of oxidation would definitely be decreased. Consequently, the difference between the rates of oxidation for the pH 8 and 6 observed will be greatly reduced. The difference could have been more pronounced if the pH in the oxidation phases has not been influenced by the IOCS.

The observed remarkable changes in the pH of the effluent of the short column during the oxidation phase were reasons to monitor the calcium content of the effluents. This is to verify the assumption that calcium carbonate might be present in IOCS. The outcome was that, the calcium content increased from 0.2 mg/l to about 4 mg/l for the oxidation tests at pH 5 and 6 indicating the presence and dissolution of calcium carbonate from the IOCS (Figure 7.7).

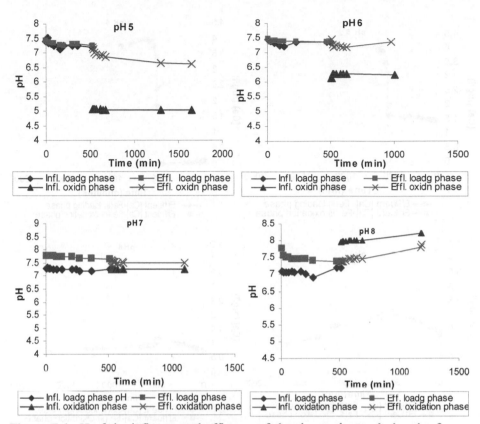

**Figure 7.6** pH of the influents and effluents of the short column, during the ferrous loading and oxidation of adsorbed Fe (II) at different feed water pH levels.
loadg = loading; Infl. = Influent; Effl. = effluent; oxidn = oxidation.

The IOCS is definitely the source of the calcium since the aerated water was not dosed with calcium. Dissolving of calcium carbonate was an option since the aerated feed water with its low pH and no calcium, was aggressive against the calcium carbonate. However, the calculated mass balance during a test run could not fully explain the observed phenomenon. More hydrogen carbonate should have been formed based on observed pH increase.

An explanation might be the presence of $FeCO_3$ and the interaction between part of the formed $H^+$ and the $FeCO_3$ (equation 7. 8). The $FeCO_3$ was possibly precipitated during the formation of the coating in the period that the IOCS was in use in the groundwater treatment plant.

$$CaCO_{3(s)} + H^+ \leftrightarrow Ca^{2+} + HCO_3^- \qquad (7.7)$$

$$FeCO_3 + H^+ \leftrightarrow Fe^{2+} + HCO_3^- \qquad (7.8)$$

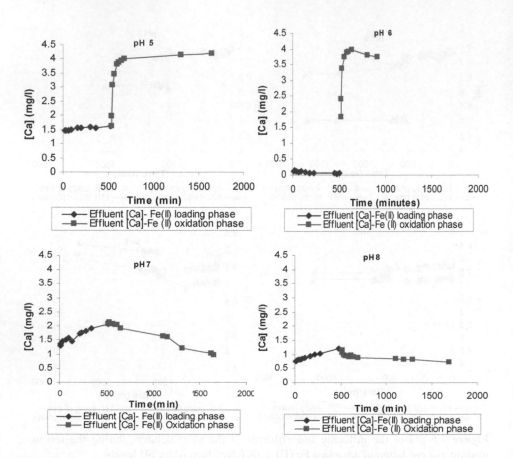

**Figure 7.7** Calcium content of the filtrates taken during the adsorbed ferrous oxidation at the various pHs. The feedwater had no calcium added to it.

The drop in pH observed for the test done at pH 8 can be explained by the formation of $H^+$ when adsorbed $Fe^{2+}$ is being oxidized (see equation 7.1). However the formed $H^+$ will certainly combine, at least partly, with calcium carbonate and ferrous carbonate in the coating and result in buffering the pH.

One other remarkable observation in Figure 7.7 is the relatively very low calcium content in the effluent during the loading prior to oxidation at pH 6. The adsorbed ferrous oxidation experiment at pH 6 was the first in the series of the oxidation experiments that were conducted. The oxidation experiment at pH 6 was therefore conducted on a fresh IOCS media and the effect of the aggressive feed water on the calcium of the IOCS was not so pronounced. The long tailing and possibility of continuous oxidation process coupled with the release of protons might have enhanced the aggressivity on the calcium of the IOCS in the subsequent oxidation experiments.

Since the interaction between the feed water during the oxidation stage of the experiment resulted in a pronounced buffering of the pH, the results are partly conclusive.

## 7.4.4 Column experiment – adsorbed $Mn^{2+}$ oxidation with Aquamandix

During the loading onto the Aquamandix, manganese present in the feedwater was almost completely adsorbed. Net amounts of 221 mg and 216 mg of manganese were adsorbed for the pH 6 and pH 8 oxidation experiments. These amounts of manganese adsorbed represent just about 4.2% of the available Aquamandix adsorption capacity for manganese. The manganese adsorption capacity of the column was theoretically calculated to be about 5200 mg Mn (II). The total solid volume of the column was 1.54 litres, so only 0.07 litres of the solid volume would be occupied by the manganese adsorbed. With 221 mg and 216 mg of $Mn^{2+}$ loaded, 64.1 $mgO_2$ and 63.0 $mgO_2$ respectively would be required for complete oxidation. The corresponding volumes of aerated feed water for these amounts of required oxygen for oxidation has been indicated on the figure 7.8 with a vertical line (i.e. 6.9 L of aerated water with DO of 9.4 $mgO_2/L$ for pH 6) and a broken line (i.e. 7.3 L of aerated water with DO of 8.6 $mgO_2/l$ for pH 8). In calculating and indicating these volumes of aerated feed water on the graph (Figure 7.8), the volume (approximately, 1 L) required for replacement of anoxic water in the filtration column has been considered.

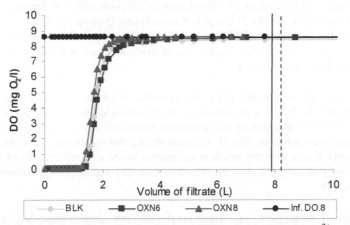

**Figure 7.8** Oxygen breakthrough curves for oxidation of adsorbed $Mn^{2+}$ at pH 6 and 8. BLK - Blank at pH 8; OXN 8 -oxidation at pH 8; OXN 6 - oxidation at pH 6 and Inf. DO. - Influent Dissolved Oxygen content at pH 8.

The oxygen breakthrough curves for the blank experiment together with the oxidation experiments at pH 6 and 8 have shown no observable difference in their oxygen uptake (Figure 7.8). Under the conditions applied and within the period (i.e. 120 minutes) of monitoring, virtually no oxidation of the adsorbed manganese occurred. This unexpected result indicates that $Mn^{2+}$ being adsorbed onto Aquamandix, does not ensure (very) fast oxidation.

In applying Aquamandix in practice it might be expected that initially the manganese removal is adequate. However, after a certain period breakthrough will occur, unless in the meantime an active catalyst has been developed through the adsorptive / oxidation process.

## 7.5    Conclusions

- Ferrous ions adsorbed onto IOCS in a short column at pH 5, 6, 7 and 8 consumed oxygen when aerated water was fed into the column. The breakthrough of oxygen in the short column filtrate appeared already after 12.8, 8.6, 30.7 and 29.1 % of the total required oxygen was consumed at pH 5, 6, 7 and 8 respectively. These results indicate that the rate of oxidation of adsorbed $Fe^{2+}$ at pH 7 and 8 is much higher than at pH 5 and 6. However the difference is much smaller than expected based on kinetics of ferrous oxidation in aqueous solution.

- This smaller difference is attributed to the phenomenon that the pH of the effluent of the column increased up to about 7.2, 7.2 and 7.5 during the oxidation phase of the adsorbed ferrous experiments using aerated feed water at pH 5, 6 and 7. This is in contrast with the expectation that the pH will drop because of the formation of $H^+$ due to oxidation of adsorbed Fe (II). This pH increase is attributed to the dissolution of calcium and / or ferrous carbonate from the IOCS. At pH 8, the pH of the effluent dropped to 7.3 – 7.7 basically due to the formation of $H^+$. Part of these $H^+$ was possibly neutralized by the IOCS. As a consequence, higher rate of oxidation was observed at pH 5, 6 and 7 and a lower rate at pH 8.

- The assumption that the pH is being buffered by the interaction of the feed water and the IOCS is supported by the observation that the ferrous adsorption onto the IOCS and subsequent oxidation proceeded with a continuous release of calcium into solution. The $H^+$ released during the oxidation of the ferrous ions probably reacts with the calcium carbonates in the mineral coating of the IOCS there by releasing the calcium. The lower the pH, the larger the buffering effect on the pH.

- Adsorbed manganese on Aquamandix did not demonstrate any uptake of oxygen within the test period of 120 minutes. Oxidation of adsorbed manganese is probably slow at least up to pH 8. This unexpected result suggests that $Mn^{2+}$ being adsorbed onto the Aquamandix, is not a guarantee that it will be oxidized rapidly. This finding makes Aquamandix attractive only as an alternate filter medium, if an adsorptive catalytic manganese oxide coating is to be formed rapidly during operation.

It is recommended that additional research should be done using artificially prepared IOCS, having no carbonates in the coating to monitor the rate of adsorbed ferrous oxidation. Furthermore, segmented column tests are recommended to monitor the uptake of oxygen under more defined and stable conditions. This is necessary because in the preliminary oxidation tests, the oxygen concentration and adsorbed Fe (II) varied in the height of the column and in time; this limited unambiguous interpretation of the results.

In addition, long term experiments need to be conducted to verify the development of adsorptive catalyst on Aquamandix under different conditions (see Chapter 5).

# 7.6    References

Aqua Techniek, (2007) Aquamandix:
    http://www.aqua-techniek.com/html/filtermedia.htm. (Last accessed - September, 2008).
Barry, R.C., Schnoor, J.L., Sulzberger, B., Sigg, L. and Stumm, W. (1994) Iron oxidation kinetics in an acidic alpine lake. *Water Research*, 28 (2), 323 – 333.
Ghosh, M. M. and O'Connor, J. T. Engelbrecht, R. S. (1996) Precipitation of iron in Aerated Groundwater: *Journal of Sanitary Engineering Division*, ASCE; vol. 90, No SA1, 199-213, paper 4687.
Standard Methods for the examination of water and wastewater (AWWA) (2005) Digestion for metals. Centennial Edition. Mary Ann H. Franson (managing editor). Port City Press, Baltimore, Maryland USA. Part 3030D pp. 3-7 to 3-10.
Sharma, S.K. 2002 Adsorptive iron removal from groundwater. PhD Thesis. UNESCO-IHE, Instistute for Water Education and Wageningen University. Swets and Zeitlinger B.V., Lisse.
Stumm, W. and Lee, G.G. (1961) Oxygenation of ferrous iron: *Industrial Engineering and Chemistry*; vol 53, No 2, 143-146.
Stumm, W. & Morgan, J.J. (1996) Aquatic Chemistry: Chemical equilibria and rates in natural waters, 3$^{rd}$ edition. Wiley, New York, USA, pp 462, 683.
Tufekci, N. and Sarikaya, H.Z. (1996) Catalytic effect of high Fe(III) concentration on Fe(II) oxidation. Water Science and Technology, 34(7 – 8), 389-396.

# Appendix 7.1

**Table 7.4** Mass balance of adsorbed ferrous oxidation.

| pH | Vol. of Fe(II) stock used (ml) | Flow rate (litres/hour) | Dissolved Oxygen content of aerated water (mg/l) | Fe(II) lost in samples used for pH and Fe(II) monitoring | Net of Amt of Fe(II) adsorbed (mg) | Expected $O_2$ to be consumed (mg) | Expected breakthrough time & volume of aerated water (min) | Expected breakthrough time & volume of aerated water (litres) | Actual amt. of $O_2$ consumed (mg) | Actual breakthrough time & volume of aerated water (min) | Actual breakthrough time & volume of aerated water (litres) | % Fe(II) oxidized |
|---|---|---|---|---|---|---|---|---|---|---|---|---|
| 5 | 2440 | 7.1 | 8.3 | 3.3 | 602.2 | 84.3 | 86.0 | 10.2 | 10.8 | 11.0 | 1.3 | 12.8 |
| 6 | 2350 | 7.0 | 8.6 | 5.5 | 582.0 | 81.5 | 81.2 | 9.5 | 7.0 | 7.0 | 0.8 | 8.6 |
| 7 | 2347 | 7.0 | 8.6 | 29.1 | 557.7 | 78.1 | 78.2 | 9.1 | 23.9 | 24.0 | 2.8 | 30.7 |
| 8 | 2597 | 7.0 | 8.3 | 5.2 | 644.0 | 90.2 | 92.7 | 10.8 | 26.3 | 27.0 | 3.2 | 29.1 |
| 7 (faster rate) | 2450 | 14.2 | 8.7 | 11.8 | 595.2 | 83.3 | 40.3 | 9.5 | 3.1 | 1.5 | 0.4 | 3.7 |

Concentration of Ferrous stock (250 mg/l)

# CHAPTER EIGHT

# SUMMARY AND CONCLUSIONS

# 8.1    Introduction

## 8.1.1 Groundwater use and quality

Groundwater abstraction and usage as raw water resource for drinking water production has been and continue to be the preferred practice in most parts of the world. Globally approximately 2 billion people rely on groundwater as the only source for drinking water. In the Netherlands, about 70% of the drinking water production has been taken from groundwater sources. In Ghana the use of ground water is not as high as in the Netherlands but still substantial considering that about 45% of the total drinking water production is groundwater. Groundwaters are generally free from fecal coliform bacterial and other microbial contamination, have low suspended particles and low turbidity. Aquifers are naturally protected by the soil and the underlying vadose zone and therefore relatively less vulnerable to anthropogenic pollution. However, as a result of large storage and long residence times when aquifers become polluted, the contamination is persistent and difficult to reverse. As a result of the reactions and interactions (like dissolution, precipitation, redox, adsorption, ion exchange, radioactivity and microbial metabolism) occurring within the geological formations of the aquifer, contaminants like manganese, iron, arsenic, ammonium, hydrogen sulphide, chromium and occasionally traces of radionuclides like uranium-238, radium-226, radon-222, etc could get mobilized into the groundwater. Some of these contaminants like the arsenic and radionuclides are important primarily from their human health standpoint in that they increase the risk of cancer when ingested. Until recently manganese was known together with iron to pose aesthetic problems in drinking water; current research indications have shown that exposure to manganese concentration beyond 0.4 mg/l in drinking water could pose a health hazard.

Presently about 1 billion people in the world lack acess to safe drinking water; the August, 2008 edition of the WHO's 'Safer water, better health' report highlights the plight of Africa with regards to the incidence of death resulting from water related diseases. In that particular report, the figures presented indicated that 15 – 20% of deaths in Africa are associated with water. Considering the population of Africa to be hovering around 922 million people and its population growth rate of 5%, such high percentage of deaths associated with water can translate into huge absolute numbers. We cannot therefore underestimate the importance of safe water (and of course hygiene) to the health of a population.

To appropriately and effectively treat groundwater for drinking water production the availability of data on the quantity, quality, hydrogeology of the aquifer, hydrological data etc are needed. However, in many parts of the world the systems for acquiring, storing, processing and managing these data are either none existent or rudimentary. In this study therefore, surveying and acquisition of water quality data with emphasis on the occurrence of manganese, iron and arsenic in some wells in Ghana became one of the focal areas of the research.

The WHO health based guideline value for manganese and arsenic in drinking water are 0.4 mg/l and 10μg/l respectively. Iron has no health based guideline value, however to avoid consumer's complaints about the aesthetic nature of the drinking water, the EC and the Netherlands directives for the concentration of iron in drinking water has been put at 0.2 and 0.05 mg/l respectively. Manganese appears in groundwater as $Mn^{2+}$, which is released under anoxic conditions, from a great variety of manganese bearing minerals e.g. $MnO_2$, $MnCO_3$, $Mn(OH)_2$, $MnOOH$, etc. Concentrations in groundwater are in the range of $0 - 10$ mg/l. Under oxic conditions, $Mn^{2+}$ is converted into manganese with higher valences and forms insoluble compounds. Iron present in groundwater mainly consists of $Fe^{2+}$ because of its high solubility. The main sources of iron are the dissolution of Fe(II) bearing minerals e.g ilmenite ($FeTiO_3$), siderite ($FeCO_3$), grunerite [$Fe_7Si_8O_{22}(OH)_2$] etc. Arsenic in groundwater originates mainly from geogenic sources and to a little extend, anthropogenic sources. Arsenic occur as a major constituent in more than 200 minerals including elemental arsenic, arsenides, arsenates, arsenites etc. Arsenic can exist in the organic and inorganic forms. Arsenic can exist in any of these valence states: $As^{3-}$, $As^0$, $As^{3+}$ and $As^{5+}$. In groundwater, the dominant form of the As is a function of pH and redox potential.

## 8.1.2 Groundwater treatment

For the removal of iron and manganese from groundwater two different process modes are applied namely:
- oxidation / flocculation mode;
- adsorptive / oxidation mode.

In the oxidation / flocculation mode, $Fe^{2+}$ is oxidized with dissolved oxygen present in the water and / or an oxidant e.g. potassium permanganate. The oxidized $Fe^{2+}$ forms the insoluble ferric hydroxide, which is removed by sedimentation and / or rapid (sand) filtration.

The adsorptive / oxidative mode makes use of adsorption of $Fe^{2+}$ on the filter media. The adsorbed $Fe^{2+}$ is oxidized by the introduced oxygen (i.e. the dissolved oxygen) and / or an oxidant, intermittently or continuously. Adsorbed and subsequently oxidized $Fe^{2+}$ forms a coating with a high adsorption capacity that grows during prolonged operation period. In the adsorption / oxidation mode, oxidation of dissolved $Fe^{2+}$ is inevitable, however can be minimized by lowering the pH and reducing the pre-oxidation time. In ensuring these latter conditions, high filtration rates and / or low backwash frequency can be achieved.

The oxidation / flocculation mode in not efective in cases where only oxygen is used as the oxidant for the oxidation of the dissolved $Mn^{2+}$ because the rate of oxidation of $Mn^{2+}$ is very low at pH values below 9. Chemical oxidation is needed. The Adsorptive / oxidation mode is effective if chemical oxidation is applied to oxidize the adsorbed $Mn^{2+}$ which then forms a coating with (probably) a high adsorption capacity. Many plants don't apply chemical oxidation but rather make use of the dissolved oxygen in the water to avoid the introduction of chemicals and its associated disadvantages e.g. cost, complexity of the process and probable formation of carcinogenic by-products.

Chemical oxidation is successfully applied at pH levels down to 7.0; indicating that the rate of oxidation of adsorbed manganese with oxygen is too low at pH levels below 9.

In situations where the groundwater has high arsenic concentration, the arsenate and arsenite ions are adsorbed onto hydrous ferric oxides. In addition, the iron oxide coating formed in the adsorptive mode has a high adsorption capacity for arsenic (III) and (IV). For this reason, iron oxide coated sand (formed in rapid sand filters removing iron from groundwater) has been selected and successfully applied as adsorbent in the UNESCO-IHE Family filter for arsenic removal.

## 8.1.3 Need for research and objectives

Treating groundwater to meet the stringent requirement for manganese, iron and arsenic in drinking water supply and reduce the operation and maintenance costs of distribution systems will require an in-depth knowledge of the source groundwater quality as well as application of more efficient removal methods. In Ghana, there is limited information on the presence of manganese, arsenic and iron in all the about 20 000 boreholes used for drinking water, most of them without any treatment. This situation is enough justification for this study focusing on the collection and analysis of well water samples from various parts of Ghana to provide additional information on the levels of manganese, arsenic, iron and other quality parameters of the groundwater used for drinking water.

In his study, Sharma (2002), revealed the mechanisms involved in adsorptive iron removal and enabling measures that further improve the iron-removing processes in practice e.g. shortening ripening time, improving filtrate quality and reducing filter backwash frequency. A similar study on manganese removal is still lacking, while there is a strong need from practice to improve this process. Frequently encountered problems are: gradual loss of manganese removal efficiency and very long start up periods when filter material has been replaced.

In addition, during the extended field tests of the UNESCO-IHE – Family filters for arsenic removal from ground water, problematic manganese removal has been observed. Initially manganese removal started up slowly; after a couple of months manganese release occurred. The manganese concentrations in the filtrate largely exceeded the concentration in the raw groundwater.

The study has the following specific objectives:
1. To screen the groundwater quality with highlights on the presence of manganese, arsenic and iron content in selected regions of the gold-belt zone of Ghana. And to identify the geological formations associated with the contaminated aquifers.
2. To determine manganese adsorption capacities of iron oxide coated sand under various process conditions.
3. To determine the effect of pH on adsorption capacities of selected media for manganese and to model the adsorption phenomenon.

4.  To study the effect of manganese and iron loading on the formation of a catalytic manganese oxide coating and the subsequent effect on the start-up of manganese removal in pilot rapid sand filters.

5.  To study the rate of adsorption of Mn(II) onto one or more selected media under different oxic conditions.

6.  To study the release of manganese from filter media and investigate measures to optimize the performance of the UNESCO-IHE family filter in the removal of manganese, arsenic and iron when treating water with high ammonium content.

7.  To investigate the oxidation kinetics of adsorbed iron and manganese at different aqueous pH values.

## 8.2     Presence of manganese, arsenic and iron in the groundwater within the gold-belt zone of Ghana

To provide additional information on groundwater quality and specifically determine the possible presence of manganese, arsenic and iron in groundwater in Ghana, nearly 290 well water samples from three regions namely Ashanti, Western and Brong-Ahafo in the gold-belt zone of Ghana, were collected and analyzed. Aside these regions, areas in the Eastern region of Ghana that are not located within the gold belt zone were also sampled to have insight into the groundwater quality outside the gold belt zones and the influence of the various geological formations on the groundwater quality. The data about the age and depth of the wells were also acquired and correlated to groundwater quality.

From the experimental analysis on the well water samples, the history of the wells and data from the geological information systems (i.e. GIS), the following conclusions were drawn:

- Unlike other areas in the world e.g. Bangladesh, Southern Taiwan etc., where high arsenic concentration is prevalent, the groundwater in the basement rock of Ghana is relatively less vulnerable to the generation of high dissolved arsenic concentrations. Arsenic presence at levels >10 µg/l was found in 7%, 12% and 5% of wells tested in the Ashanti, Western and Brong-Ahafo regions, respectively. Of all the samples analyzed, the Anyinase community (Western region) groundwater possessed the highest arsenic contamination of 120 µg/l. It is estimated that between 500 000 and 800 000 inhabitants in the communities covered in this study use untreated water with [As] > 10 µg/l. About 70% of the wells with arsenic concentration > 10 µg/l in the regions had been in use for more than 15 years.

- Manganese levels in 13% and 29% of the wells in Ashanti and Western regions exceeded the WHO health-based guideline of 0.4 mg/l.

- Thirty five percent, 50% and 5% of the sampled wells in Ashanti, Western and Brong -Ahafo regions, respectively had iron content >0.3 mg/l.

- The communities with high arsenic contamination are situated on the Birimian sediments, Birimian volcanics and Tarkwaian formation in the gold-belt zone.

- All the communities with higher arsenic contamination (>50 μg/l) have iron levels >1.0 mg/l and manganese >0.4 mg/l. Most of them have pH within the range of 5.6 to 6.5. The release of arsenic into the aquifers is presumably due to arsenopyrite oxidation and reductive dissolution and desorption of ferric oxides. A linear multiple regression showed significant correlation between the concentration of arsenic with acidic pH, Fe and Mn content of the arsenic contaminated well water. There is however no significant correlation between arsenic concentration and the age of the contaminated well.

- Most of sampled wells in the Eastern region, outside the gold-belt zone had manganese >0.4 mg/l and iron >0.3 mg/l but no arsenic contamination. The wells sampled in the Eastern region are located within the Upper Voltain and Togo series geological formations that have no gold deposits. These formations have different characteristic features from the Birimian and Tarkwain formations that are often associated with arsenopyrites. This shows that geology has a major impact on the occurrence of arsenic in groundwater.

## 8.3    Adsorptive removal of manganese (II) from the aqueous phase using iron oxide coated sand

Iron oxide coated sand (IOCS), a by-product from the rapid sand filters of groundwater treatment plants has been found to be a potential adsorbent for metallic ions e.g. manganese, iron, arsenic, lead etc. The effect of process conditions and the mechanism of Mn (II) adsorption on IOCS, however, haven't been thoroughly investigated. The determination of the effects of the process conditions will give a better understanding of the adsorptive manganese removal process and associated mechanisms. The focus of this study was therefore directed towards determining the capacity, rate, mechanisms involved and the effect of process conditions on the adsorption of Mn (II) onto IOCS using laboratory scale batch reactors with modeled water. In addition, kinetic studies using the Linear Driving force, Lagergren and Potential Driving Second Order Kinetic (PDSOK) models were conducted to examine the rate of manganese (II) adsorption onto aggregate and pulverised IOCS.

Based on the experimental results the following conclusions were drawn:

- Alkalinity and pH have a marked effect on solubility of Mn (II). This solubility is governed by the formation of manganese carbonate. Manganese hydroxide has a much higher solubility. At pH values of 8 and higher, calculated solubility of Mn (II) is very limited (1 – 2 mg/l or lower) even at low alkalinity (60 ppm).

- IOCS demonstrated a substantial adsorption capacity for manganese (II) under anoxic and oxic conditions. Adsorption data fit reasonably well the Freundlich isotherm model. Adsorption capacities ('K' values increase from 0.0024 to 147) increase with pH from 5 to 8 and remarkably, in the range of pH 7 to 8.

- The rate of adsorption of Mn (II) on IOCS at pH 6 is identical under oxic and anoxic conditions. Therefore there is no indication that under oxic conditions at pH 6, adsorbed Mn (II) is oxidised and additional adsorption capacity formed.

- Plots of the kinetic models, i.e. Linear driving force, Lagergren and Potential Driving Second Order Kinetic models show a reasonable linearity. However, for aggregate IOCS, the initial slopes decreased with progressing adsorption. This phenomenon might be attributed to the presence of easily accessible and less accessible adsorption sites (e.g. in narrow pores) and / or decreasing pH in the pores as a result of Mn (II) adsorption on IOCS.

- Predicting the equilibrium concentration making use of three kinetic models turned out to be not a feasible option due to changing adsorption rate constants in the course of the adsorption process.

## 8.4    Manganese adsorption characteristics of selected filter media for groundwater treatment: equilibrium and kinetics

Aside the aesthetic problems that the presence of manganese causes in drinking water, chronic exposure of consumers to levels beyond 0.4 mg/l may pose a health hazard. Many ground water treatment plants applying the autocatalytic-adsorptive process in manganese removal use sand as the media in their rapid sand filters. To optimize manganese removal in filters there is the need to look for other readily available media that may have a relatively higher manganese adsorption capacity. To address this issue, laboratory scale batch adsorption experiments were designed and conducted using modeled water to determine the manganese adsorption capacities at different process conditions for eight filter media including:

1.  Aquamandix (a commercial product);
2.  IOCS (two types of this by-product originating from groundwater treatment plants of Noord Bargeres and Spannenbroog in the Netherlands);
3.  Iron-ore (a virgin material);
4.  Laterite (a virgin material);
5.  manganese green sand (a commercial material – in its original and regenerated forms);
6.  Quartz sand (a virgin material).

The experimental data obtained in the batch manganese adsorption experiments of these filter material were fitted to the Potential Driving Second Order Kinetic (PDSOK), Lagergren and the Dubinin-Kaganer-Radushkevisch (DKR) models. Based upon the experimental results, the following conclusions were made:

- The Aquamandix, IOCS, LmIOCS, Iron-ore and Laterite demonstrated the potential to adsorb manganese at pH 6 and 8 under both oxic and anoxic conditions. Manganese green sand showed manganese adsorption capacity only at pH 8. The manganese adsorption capacities of all the media tested were much higher at pH 8 than pH 6.

- The manganese adsorption capacities of the various media were found to be in the following order:

  At pH 6: AQM > IOCS > LmIOCS > (Fe-ore) > LAT > Virgin sand;
  At pH 8: AQM > IOCS > LmIOCS > RMGS > Fe-ore > MGS > LAT > Virgin sand.

  Aquamandix demonstrated the highest adsorption capacity and much higher than sand. Consequently Aquamandix qualifies as a potential alternative for the sand that is commonly used in manganese removal filters in practice.

- Iron oxide coated sand, containing sufficient manganese is the second best option with a reasonably high adsorption capacity and being a by-product of the water treatment plants could be the cheapest substitute. Virgin sand has by far the lowest adsorption capacity.

- The adsorption potential of Aquamandix, iron oxide coated sand (with high and low manganese) and manganese green sand, coincides with manganese content indicating that manganese is essential for adsorption capacity of media.

- The adsorption kinetics of manganese under anoxic condition on different media gave the best fit with the PDSOK model, followed by the DRK model. The Largergren adsorption process is best described by the PDSOK model.

- The linear relationship obtained for the PDSOK model and the high sorption energy obtained from the DKR model indicate that the manganese adsorption onto the Aquamandix and IOCS follows chemi-sorption.

- Aquamandix media demonstrated autocatalytic adsorption under oxic conditions and pH 8. Results obtained with other media were less pronounced and need to be studied more in detail.

## 8.5    Manganese removal from groundwater; problems in practice and potential solution

Most drinking water production plants in Europe, like the Brabant water company in Haaren, the Netherlands that use rapid sand filters for the removal of manganese from groundwater normally experience long ripening and start-up of manganese removal on newly installed sand media. The start-up of the manganese removal has been found to take several weeks till months to be established. Reducing this period in order to prevent the loss of water during this phase has become an issue of concern.

To devise measures to solve this problem, pilot and bench scale experiments were conducted to investigate the mechanism, influence of operational conditions and measures that enhance the development of adsorptive / catalytic coating on the sand media and thus enhance the manganese removal capacity of the media. For the pilot short column experiments, three sand filters were fed with filtrate with low iron and ammonium contents (0.0.4 mgFe$^{2+}$/l and 0.55 mgNH$_4^+$/l) from the Haaren Ground Water Treatment Plant (GWTP). In two of the short column, the Haaren GWTP filtrate (i.e. the influent of these two filters) was dosed with manganese and fed to the two filters at different filtration rates (0.6 m/hr and 2.8 m/hr). The third filter had no manganese dosing in its influent (i.e. the Haaren GTWP filtrate). The third filter was run at a filtration rate of 2.8 m/hr).

After several weeks of filter runs, bench scale experiments were conducted using the coated sand media from the short columns used in the pilot experiments to determine their manganese removal potentials under both oxic and anoxic conditions.
From both the pilot and the bench scale experimental results the following conclusions were drawn:

- The three filters filled with virgin quartz sand developed full manganese removal only after several weeks of operation. The filters fed with water containing 0.6 mg Mn(II)/l and operating at filtration rates of 2.8 and 0.6 m/h developed complete manganese removal in 79 and 51 days of operation respectively. The filter fed with 0.04 mgMn(II)/l and running at 2.8 m/h achieved full manganese removal only after 100 days of operation

- The positive effect of a low rate of filtration is attributed to the less frequent backwashing of the filter run at lower filtration rate.

- The long period that the filter fed with low manganese needed to arrive at full manganese removal is attributed to the low influx of manganese required to develop sufficient adsorptive catalyst surface area.

- Nitrite might have a negative effect on the development of the adsorptive catalyst however this has to be verified.

- Adsorption capacity measurements, measured under oxic conditions on the media of the three filters, taken after 15 weeks of pilot operation, demonstrated a continuous removal of manganese. Conducted batch adsorption experiments are a potentially powerful tool in judging the removal capacity of manganese removing filtering materials. Standardization of the procedure is however required.

- The measured removal rate capacity of the top layer of the media after 15 weeks of operation, determined in laboratory and bench scale tests, was too low to explain the observed complete removal of manganese in the filters fed with 0.6 mg Mn(II) /l. The manganese adsorptive catalyst on media of the top layers might have been partly covered with ferric hydroxide and / or the catalyst lost a part of its manganese removal capacity due to aging in the period between sampling and measuring.

- Microscopic photographs of the sand grains surface of the sand indicate that the catalyst does not develop as a homogenous layer, but develops in the form of specific spots.

- The spectra obtained from Scanning Electron Microscopy investigations on the surface of coated sand grains taken from the top layer of the filters confirmed the non-uniform development of the catalyst on the surface coating.

- Out of the six filtration media, a manganese bearing mineral (Aquamandix) demonstrated the highest adsorptive capacity measured under anoxic conditions. As its capacity is much higher than sand, Aquamandix is a potential candidate to substitute (or partially substitute) sand in situations of a slow start-up of manganese removal.

## 8.6    Optimising the removal of manganese in UNESCO-IHE arsenic removal family filter treating groundwater with high arsenic, manganese, ammonium and iron

On-going research at UNESCO-IHE has shown over the years that IOCS can be effectively used to remove arsenic, iron, chromium etc. from groundwater and modeled water with high contents of these contaminants. These experiments have also shown that the IOCS can be regenerated to restore its adsorptive capacity after exhaustion. However, the studies showed poor performance of UNESCO-IHE family filter (using IOCS) in terms of Mn removal ranging from no removal at all to increased concentration of manganese in the filtrate when high amounts of $NH_4^+$ occur in the ground water. In order to investigate the manganese release phenomenon and optimize the performance of the UNESCO-IHE family filter with focus on manganese removal, batch and short column experiments conducted under laboratory condition using IOCS and Aquamandix. Different UNESCO- IHE family filters were run with i) IOCS, ii) Aquamandix layer on top of IOCS and iii) post sand filter with local sand to optimise the filters in terms of manganese removal efficiency. In running the columns, model water containing high contents of manganese (1 mg/l), arsenic (200 µg/l), iron (5 mg/l) and ammonium (4 mg/l) were used.

The conclusions drawn from the experimental results include the following:

- The filter filled with IOCS (Family filter) treating model water with high ammonium concentration, removed initially manganese adequately. After about 10 days, manganese breakthrough occurred and subsequently manganese was released from the IOCS. These observed phenomena are attributed to the oxidation of ammonium by Nitroso bacteria, resulting in an anoxic condition in the filter. Under that condition manganese was not removed adequately; on the longer term, accumulated manganese was released, likely due to reduction of manganese dioxide to manganese (II).

- The Family filter having a polishing layer of a manganese mineral (Aquamandix) on top, removed consistently manganese below the 0.4 mg/l WHO guideline. Adequate removal of manganese is attributed to the high adsorption capacity for manganese (II) of the polishing layer.

- Post sand filtration after aeration did not remove manganese from the effluent of the Family filter within seven weeks operation at pH 6.8. Manganese removal started after this period indicating that the formation of catalytic manganese oxides at about pH 7 is very slow. This is in concert with the results in Chapter 5 and findings in practice.

- A Family filter with a sand layer on top of IOCS fed with model water without ammonium, showed significant higher manganese removal efficiency. This result demonstrates that the presence of ammonium reduced the removal of manganese and caused the release of manganese by creating anoxic conditions, due to bacterial oxidation of ammonium.

- High ammonium content (4 mg/l) of the model water caused fast depletion of dissolved oxygen within both filters (i.e. Family filters 1 and 2) and the establishment of anoxic conditions with the attendant high nitrite concentrations exceeding 0.2 mg/l (WHO guideline), within the filtrates. The formation of nitrite and the inability of the filters to convert it to nitrate is attributed to the absence of sufficient oxygen in the feed water. A polishing filter fed with aerated effluent removed the nitrite adequately.

- Arsenic removal in the Family filter was initially adequate; however, soon its performance reduced. Frequently, the arsenic concentration in the effluent exceeded the level of 10 µg/l (WHO guideline). The observed deterioration of the effluent quality is attributed to the escape of ferric hydroxide flocs. These flocs contained relatively high amounts of arsenic. This dominant cause of deterioration of the effluent quality is clearly illustrated by the facts that:

  - a polishing layer of sand placed on top of the Family filter removed the arsenic to the same low levels (below 10 µg/l);
  - treatment of aerated effluent of the Family filter had the same effect;
  - filtration of the effluent of the Family filter through a membrane filter (0.45 µm) reduced the arsenic concentration substantially.

- The Family filter equipped with a polishing layer of Aquamandix on top removed arsenic (95 – 99%) consistently below 10µg/l throughout the experimental period. These results are due to the ferric hydroxide flocs and adsorption of arsenic by the Aquamandix polishing layer.

- Iron removal in the Family filter was initially adequate, however after several weeks the concentration exceeded 0.3 mg/l. Treating the aerated effluent with a filter filled with sand reduced the iron concentration adequately. The same effect was achieved by placing a polishing layer of sand on top of the IOCS in the Family filter.

- The Family filter with addition of aquamandix layer on the top of IOCS showed adequate removal of arsenic, iron, and manganese when treating model water with high $NH_4^+$. A polishing, post-sand filter will be required to remove nitrite.

- A filter filled with fine grains of Aquamandix, treating the aerated effluent of a Family filter with IOCS only, is recommended to ensure adequate removal of manganese and nitrite. In addition this post treatment will improve the removal of arsenic and iron.

## 8.7 Oxidation of adsorbed ferrous and manganese: kinetics and influence of process conditions

The presence and amount of catalytic manganese and iron oxides in the composition or mineral coating of filter media contribute immensely to the removal of dissolved manganese and iron in water. Most GWTP apply aeration followed with rapid (sand) filtration to remove dissolved manganese and iron in groundwater. The adsorption process has been studied at different pH values and on different media under anoxic conditions in previous studies. However the oxidation of the adsorbed $Fe^{2+}$ and /or $Mn^{2+}$ on the filter media has not been thoroughly studied. A better understanding of the oxidation of the adsorbed $Fe^{2+}$ and / $Mn^{2+}$ phenomenon will help implement measures that will enhance continuous regeneration of the catalytic oxides. In so doing removal of $Fe^{2+}$ and $Mn^{2+}$ could be optimized.

The $Fe^{2+}$ in aqueous solutions is rapidly oxidized at pH value of 5 to 8. In contrast $Mn^{2+}$ is not oxidized in this pH range at all. This oxidation trend is what prevails in aqueous solutions; in the case of the adsorbed manganese and iron on filter media the oxidation phenomenon remains an unexplored area. To investigate the oxidation of adsorbed iron and manganese, short column experiments were run using IOCS and Aquamandix respectively under different process conditions. Model water containing iron / manganese were run through filters under anoxic conditions and subsequently oxidised using aerated water at various pHs. The rate of consumption of dissolved oxygen in the aerated influent was then monitored to determine the rate of oxidation of the adsorbed iron / manganese.

From the experimental results the following conclusions were drawn:

- Ferrous ions adsorbed onto IOCS in a short column at pH 6 and 8 demonstrated consumption of oxygen when aerated water was fed into the column. The breakthrough period for oxygen consumption by the adsorbed ferrous increased about four fold upon increasing pH from 6 to 8. 8.6% and 29.2% of the total required oxygen consumption at pH 6 and 8 respectively was consumed at the moment breakthrough appeared. This result indicates that the rate of oxidation of adsorbed $Fe^{2+}$ at pH 8 is much higher than at pH 6.

- During the oxidation phase of the adsorbed ferrous experiments using aerated feed water at pH 6, the pH of the effluent of the column increased up to 7.2 while a drop in pH was expected because the oxidation process normally results in the formation of $H^+$. This pH increase is attributed to the dissolution of calcium and / or ferrous carbonate from the IOCS. At pH 8, the pH of the effluent dropped to 7.3 – 7.7 basically due to formation of $H^+$. Part of these $H^+$ was possibly neutralized by the IOCS. These effects on pH resulted in higher rate of oxidation at pH 6 and a lower rate at pH 8.

- The hypothesis that the pH is being buffered by the interaction of the feed water and the IOCS is supported by the observation that the ferrous adsorption onto the IOCS and subsequent oxidation proceeded with a continuous release of calcium into solution. The $H^+$ released during the oxidation of the ferrous ions probably reacts with the calcium carbonates in the mineral coating of the IOCS there by releasing the calcium. At the lower pH 6, the interaction between the aerated feed water was probably more pronounced because of the higher release of calcium into the filtrate.

- Adsorbed manganese on Aquamandix did not demonstrate any uptake of oxygen. Oxidation of adsorbed manganese is probably slow at pH values up to pH 8. This unexpected result suggests that $Mn^{2+}$ being adsorbed onto the Aquamandix, is not a guarantee that it will be oxidized rapidly.

## 8.8 Reference

Sharma, S.K. 2002 Adsorptive iron removal from groundwater. PhD Thesis. UNESCO-IHE, Instistute for Water Education and Wageningen University. Swets and Zeitlinger B.V., Lisse.

## 8.9 Recommendations for further studies

1. It is recommended that a country wide inventory of the water quality of existing wells in Ghana be made and further investigations carried out with the bid to establish possible link between arsenic occurrence and the health status of

consumers. A more detailed study on the arsenic mobilization within the depth profile of the aquifers in the arsenic contaminated areas should be carried out.

2. Using GIS and GPS systems, elaborate geological survey of the other regions of Ghana with regards to occurrence of arsenic, chromium, fluoride and other inorganic contaminants in aquifers / wells used for drinking water production should be conducted and the mobilization mechanisms elucidated.

3. The capacity of IOCS to adsorb ammonium must be investigated.

4. Further studies on the oxidation kinetics of adsorbed iron focusing on the use of various filter media without any calcium carbonate impregnation would shed more light on the rate catalytic iron oxidation. Furthermore, segmented column tests are recommended to monitor the uptake of oxygen under more defined and stable conditions.

5. Long term performance of family filters using combinations of IOCS and Aquamandix as filter to treat groundwater with high arsenic, iron and ammonium and methane is recommended.

6. Several literature reports have indicated the efficiency of biomass mediated manganese removal systems. A Comparative study on manganese removal using media with high manganese removal capacity like Aquamandix and biomass mediated manganese phenomenon should carried out with the intention of designing and developing a system with a sustained high manganese removal potential.

# SAMENVATTING

Betrouwbaar drinkwater is essentieel voor het verbeteren van de gezondheid en levenstandaard van mensen en zal uiteindelijk de armoede in de wereld verlichten. Ziektes veroorzaakt door verontreinigd water zijn namelijk verantwoordelijk voor hoge sterftecijfers in veel landelijke en semi-stedelijke gebieden in de wereld. Ongeveer 20% van de sterfgevallen in Afrika kunnen hier aan worden toegeschreven. Een dergelijke hoog percentage heeft een direct effect op de inzet van de beroepsbevolking, de productiviteit in de industrie en de landbouw, en legt extra druk op financiële middelen van veel landen. Om deze redenen vraagt de aanwezigheid van verontreinigingen in het drinkwater zoals arsenicum en mangaan, steeds meer de aandacht van de waterleidingbedrijven en de autoriteiten. Hoge concentraties mangaan ( > 0.4 mg/l ) en arsenicum ( > 0.01 mg/l ) in het drinkwater kunnen een duidelijke invloed hebben op de volksgezondheid. Hoge concentraties arsenicum kunnen bijvoorbeeld verschillende soorten kanker veroorzaken wanneer het drinkwater langdurig wordt gebruikt. Chronische blootstelling aan een te hoog gehalte mangaan kan het zenuwstelsel aantasten, vergelijkbaar met de ziekte van Parkinson. Bovendien veroorzaakt de aanwezigheid van mangaan samen met ijzerdeeltjes voor een onaangename smaak van het drinkwater en laat het vlekken achter op wasgoed en sanitair.

De levering van betrouwbaar drinkwater vergt een gedegen kennis van de hoeveelheid en kwaliteit van de beschikbare waterbronnen gecombineerd met een degelijk ontwerp van de zuiveringsinstallaties, dat gebaseerd is op de nieuwste technologische inzichten. Deze inzichten zijn noodzakelijk om te kunnen garanderen dat schadelijke en andere ongewenste verontreinigingen afdoende worden verwijderd. Uit verschillende studies naar de aanwezigheid van geschikte waterbronnen blijkt dat in veel delen van de wereld hydrologische en hydrogeologische gegevens ontbreken.

Momenteel is ongeveer 45% van de totale drinkwater productie in Ghana gebaseerd op grondwater. Bestudering van de beschikbare gegevens over de kwaliteit van het grondwater in Ghana leert dat er in bepaalde gebieden, grondwater wordt gebruikt als drinkwater, dat een potentieel gevaar is voor de volksgezondheid door de hoge concentraties mangaan, arsenicum, ijzer, fluoride en nitraat. Hoge concentraties arsenicum in het grondwater komen vooral voor in Obuasi, een stad in de Ashanti regio, in het zuidelijke deel van Ghana. Dit gedeelte van Ghana is vooral bekend vanwege de goudmijnen die zich daar bevinden (EAWAG 2000). Er is echter maar weinig informatie bekend over de aanwezigheid van arsenicum in het grondwater in andere delen van het land. Reden hiertoe is kennelijk, een zeer beperkt onderzoek programma naar de aanwezigheid van arsenicum in grondwater. Bij het mobiliseren van arsenicum in grondwater spelen de aanwezigheid hiervan in het bodemmateriaal, de pH-waarde, de redox potentiaal en de aanwezigheid van ijzer en mangaan een belangrijke rol. De beschikbare informatie over de samenstelling van het water in de verschillende watervoerende lagen, in het land is zeer beperkt. Deze situatie rechtvaardigt het voorgestelde onderzoek, om monsters te nemen van drinkwater putten in verschillende delen van Ghana en hier in de niveau`s van mangaan, arsenicum, ijzer en een beperkt aantal andere parameters te bepalen.

Narmate de welvaart van consumenten stijgt, verwachten zij niet alleen dat de gezondheidsrisico's door de aanwezigheid van schadelijke stoffen tot het uiterste worden beperkt, maar ook dat het geleverde water aan esthetische eisen voldoet. Er zijn dan ook stringentere waterkwaliteiteisen voorgesteld, wat er toe leidt dat waterleidingbedrijven er toe worden aangezet verbeterde en innovatieve technologieën te gaan toepassen. Sinds de jaren negentig voert UNESCO-IHE onderzoek uit naar de verbetering van de verwijdering van ijzer en arsenicum uit grondwater. Beluchting gevolgd door snelfiltratie wordt algemeen in de meeste zuiveringsinstallaties gebruikt voor de verwijdering van deze verontreinigingen. Sharma (2002) heeft in zijn studie naar het adsorptief verwijderen van ijzer uit het grondwater de mechanismen die in dit proces een rol spelen in kaart gebracht en aangetoond dat er in de praktijk nog veel kan worden verbeterd, zoals het verkorten van de inlooptijd van de snelfilters na spoeling, de kwaliteit van het filtraat en het verlagen van de spoelfrequentie. Een vergelijkbare studie naar het verwijderen van mangaan ontbreekt tot nu toe, terwijl er in de praktijk een duidelijke behoefte is om dit proces te verbeteren. Veel voorkomende problemen in de praktijk zijn:

- de geleidelijke afname van de verwijdering efficiëntie van het mangaan in de loop van de tijd;
- de lange inwerkingtijd na vervanging van het filtermateriaal van de snelfilters, voordat volledige mangaanverwijdering wordt bereikt.

Verder is bij de uitgebreide beproeving van de UNESCO-IHE" Family filters", die bedoeld zijn voor het verwijderen van arseen uit grondwater, gebleken dat de verwijdering van mangaan problematisch kan zijn. De concrete doelstellingen van dit onderzoek zijn gezien het bovenstaande als volgt geformuleerd:

1. Het onderzoeken van grondwaterkwaliteit in Ghana waarbij de aandacht zich mede richt op de regio waarin goud wordt gewonnen. De aandacht is hierbij vooral gericht op aanwezigheid van mangaan, arsenicum en ijzer in het grondwater, en het identificeren van de geologische formaties die in verband kunnen worden gebracht met aanwezigheid van deze verontreinigingen..

2. Het vaststellen van de mangaan adsorptiecapaciteit van filterzand, dat is gebruikt voor de behandeling van ijzerhoudend grondwater en bedekt is met een laag ijzerhydroxide (coating).

3. Het bepalen van het effect van de pH-waarde op de adsorptie capaciteit voor mangaan van verschillende filter media en de modellering van het verloop van het adsorptieproces.

4. Het bestuderen van de adsorptiesnelheid van mangaan op verschillende filter media onder zuurstofhoudende en zuurstofloze condities.

5. Het bestuderen van het effect van de aanwezigheid mangaan en ijzer in het te behandelen water op de vorming van een katalytisch mangaan oxide coating en het effect hiervan op het verloop van mangaan verwijdering bij ingebruikname van nieuw filtermateriaal.

6. Het bestuderen van de afgifte van mangaan door filter media en het onderzoeken van mogelijkheden om de werking van het UNESCO-IHE filter te

optimaliseren ten aanzien van de verwijdering van mangaan, arseen en ijzer bij de behandeling van water met een hoog ammonium gehalte.
7. Het onderzoeken van de snelheid waarmee ijzer (II) en mangaan(II), dat is geadsorbeerd aan "met ijzer gecoat filterzand" en Aquamandix, wordt geoxideerd in zuurstof houdend water bij verschillende pH-waarden.

Om aanvullende informatie te verkrijgen over de kwaliteit van het grondwater in Ghana, zijn er ongeveer 290 grondwater monsters genomen uit 3 verschillende regio's te weten: Ashanti, the Western en Brong- Ahafo. Deze watermonsters zijn vervolgens geanalyseerd op de aanwezigheid van onder ander mangaan, arsenicum en ijzer. Ongeveer 13 % van de putten in Ashanti en 29% van de bronnen in de Western regio overschreden de grenswaarde van 0.4 mg/l, die door de World Health Organisation als richtlijn voor de aanwezigheid van mangaan wordt aangegeven. In Brong-Ahafo, the Western regio en Ashanti blijken van de onderzochte putten respectievelijk 5%, 25% en 50%, ijzergehalten te hebben die boven de in Ghana geldende drinkwater richtlijn voor ijzer ( 0.3 mg/l ) liggen. Er is vastgesteld dat 5 – 12% van de bemonsterde putten een arsenicum gehalte heeft, dat de richtlijn van 10 ug/l van de World Health Organisation, overschrijdt. Geschat wordt dat ongeveer 500.000 – 800.000 mensen dit ongezuiverde water gebruiken als drinkwater. Gemeenschappen binnen het onderzochte gebied, waar een hoog arsenicum gehalte in het water is aangetoond, liggen in de Birimian en Tarkwaian geologische formaties. De meeste met arsenicum verontreinigde putten (70%) zijn ten minste 15 jaar in gebruik.

Bij het onderzoek naar de adsorptiecapaciteit van "met ijzer gecoat filterzand" voor mangaan is gebleken dat de adsorptie capaciteit van bij toenemende pH-waarden sterk wordt verhoogd. Zo zijn de K-waarden respectievelijk 4,7 en 147 voor de Freundlich isothermen bij pH-waarden van 6 en 8. Bij een pH-waarde van 6 zijn vergelijkbare waarden zijn gemeten onder zowel zuurstofhoudende als zuurstofloze condities. Dit geeft aan dat er geen significante hoeveelheden geadsorbeerd mangaan worden geoxideerd bij een pH-waarde van 6, die kunnen leiden tot een extra adsorptiecapaciteit, althans binnen de periode die experimenten duurden. Er is berekend dat het waterstof carbonaatgehalte en pH-waarden van 8 en hoger, in hoge mate de oplosbaarheid van mangaan - dat wordt beheerst door mangaancarbonaat - bepalen. De oplosbaarheid is dan namelijk zeer beperkt (1-2 mg/l of lager) zelfs bij lage waterstof carbonaatgehalten (60 mg/l).

Onderzoek naar de kinetiek van de adsorptie van mangaan met de Linear Driving Force, Lagergren en Potential Driving Second Order Kinetic (PDSOK) modellen leerde dat de snelheid waarmee mangaan (II) adsorbeert, na korte tijd duidelijk afneemt. Dit verschijnsel wordt mogelijk veroorzaakt, doordat eerst de relatief gemakkelijk toegankelijk adsorptie plaatsen worden bezet, die zich nabij het oppervlak van de korrels bevinden, gevolgd door de minder toegankelijke en dieper in de poriën liggende plaatsen en/of de pH daling in de poriën van de korrels, tengevolge van adsorptie van mangaan. De veranderende adsorptiesnelheid constanten tijdens het adsorptieproces verhinderen, dat met kortdurende proeven en toepassing van een of meer modellen, de evenwichtsconcentraties kunnen worden voorspeld.

De adsorptie capaciteit van verschillende filtermedia voor mangaan zijn onderzocht in ladingsgewijze experimenten. De resultaten geven de volgende volgorde van deze capaciteiten bij een pH-waarde van 8 aan: Aquamandix > "met ijzerhydroxide gecoat filterzand" > ijzererts > mangaan "green sand" > Lateriet > nieuw filterzand. De verkregen resultaten worden voor de meeste filter media, goed beschreven met de vergelijkingen voor de Freundlich en Langmuir isothermen. De adsorptie capaciteiten waren bij een pH-waarde van 8, significant hoger dan bij een pH-waarde van 6. De meetresultaten bij pH 8, geven aan dat er sprake is van autokatalytische oxidatie van geadsorbeerd mangaan.

Drie kinetische modellen zijn toegepast om de adsorptiesnelheid/ verwijdering van mangaan in ladingsgewijze experimenten te beschrijven. Het PDSOK model gaf de beschrijving, gevolgd door het Dubinin-Kaganer-Radushkevisch model. Het Lagergren model voldeed duidelijk het minst.

Het effect van de belading van proeffilters met opgelost mangaan en ijzer op de vorming van een katalytisch werkend mangaan oxide coating en de hieruit af te leiden consequenties voor het op gang brengen van de ontmanganing in snelfilters, is onderzocht met proeffilters. Deze proeffilters zijn continu gevoed met ijzer en ammonium houdend grondwater, waaraan verschillende hoeveelheden mangaan werden gedoseerd. De ontwikkeling van de adsorberende/katalytisch werkende coating op het filterzand in de proefinstallatie was zeer traag, ondanks de relatief hoge pH-waarde van 8. Lage mangaan concentraties en frequente spoeling resulteerde in een trage opstart van de mangaan verwijdering. Het kan niet worden uitgesloten dat ook nitriet hierbij een negatief effect heeft. Proeven uitgevoerd met het filterzand uit de proefinstallatie,waarop zich een coating had ontwikkeld, laten zien dat de adsorptiesnelheid/ oxidatie snelheid van mangaan aan materiaal afkomstig uit de bovenste laag van het filterbed te laag is om de volledige verwijdering van mangaan te kunnen verklaren. Waarschijnlijk is de adsorptiecapaciteit/katalytische capaciteit van de bovenste laag van het filterbed verminderd doordat deze gedeeltelijk bedekt is met ijzer hydroxide.

Om de afgifte van mangaan door de filtermedia te onderzoeken en maatregelen te ontwikkelen om de mangaan verwijdering van de filters te optimaliseren zijn experimenten uitgevoerd met verschillende varianten van de UNESCO-IHE "Family filters" die gevuld zijn met: i) "met ijzerhydroxide gecoat filterzand" (Filter 1) ii) een laag Aquamandix op het gecoate filterzand (Filter 2) iii) filterzand en toegepast als nafilter (Filter 3).Bij deze experimenten zijn de filters gevoed met water met relatief hoge concentraties mangaan (1 mg/l), arsenicum (200 µg/l), ijzer (5mg/l) en ammonium (4 mg/l). In Filter 1, dat gevuld is met filterzand met een coating, vertoonde aanvankelijk afdoende verwijdering van mangaan, vrij spoedig veranderde dit en nam het mangaangehalte in het filtraat sterk. De arsenicum- en ijzer verwijdering bedroeg de eerste week 94-99% maar daalde daarna tot 70-95% . Filter 2, uitgerust met laag Aquamandix op het "met ijzer gecoate filterzand" verwijderde consistent arsenicum (95-99%) en mangaan (90-100%) tot onder het niveau van de WHO richtlijn van respectievelijk 0.01mg/l en 0.4 mg/l.. Ook ijzer werd consistent verwijderd (95-100%) tot onder de norm van 0.3 mg/l. Het filtraat van Filter 2 bevatte echter een hoog nitrietgehalte (gemiddeld 2.2 mg /l). Filter 3, gevuld met nieuw filter zand, was in serie geplaatst met Filter 1 en deed dus dienst als nafilter, verwijderderde arsenicum, ammonium en nitriet tot onder de WHO richtlijn. Dit filter verwijderde ijzer tot niveaus lager dan 0.3 mg/l. De nitriet concentraties in het filtraat van filter 3 waren echter hoog te weten gemiddeld 2.15 mg/l. Na 7 weken, werd bij een pH-waarde van 6.8 nog steeds mangaan verwijdering waargenomen. Deze waarneming geeft aan dat de vorming van adsorptieve/katalytische mangaanoxiden bij deze pH-waarde traag verloopt, dan wel afwezig is . De toevoer en/of het niveau van mangaan speelt een rol in de ontwikkeling van mangaanoxiden op het oppervlak van het filterzand, de pH-waarde is echter de meest bepalende factor. Een laag fijn filterzand geplaatst op het met ijzer gecoate filterzand verwijderde de ijzervlokken met daaraan gehecht arsenicum. Het hoge ammonium (4 mg/l) gehalte in het voedingwater veroorzaakte in korte tijd een volledig verbruik van de aanwezige zuurstof en creëerde zuurstofloze omstandigheden met daarmee gepaard gaande hoge nitriet concentraties in de filtraten. De belangrijkste oorzaak van de hoge nitriet concentraties is waarschijnlijk het ontbreken van voldoende zuurstof voor de omzetting van nitriet naar nitraat en de het niet overleving van de Nitrobacter bacteriën. Het UNESCO-IHE "Family filter" met een laag van een Aquamandix op het "met ijzer gecoate filterzand", bleek de beste resultaten te geven door arsenicum, mangaan en ijzer afdoende te verwijderen. Een nafilter gevuld met zand kan bij deze opzet nodig zijn om nitriet afdoende om te zetten in nitraat.

De oxidatie van het adsorbeerde $Fe2+$ en/of $Mn2+$ op filtermedia speelt een belangrijke rol bij de verwijdering van opgelost ijzer en mangaan in grondwater. $Fe2+$ in water opgelost wordt snel door zuurstof geoxideerd bij pH-waarden vanaf 5. Echter $Mn2+$ wordt bij pH-waarden beneden 9 in het geheel niet geoxideerd. In de praktijk wordt mangaan verwijderd in snelfilters bij pH-waarden hoger dan 6.9. Dit verschijnsel geeft aan dat adsorptie van $Mn2+$ op filtermedia de oxidatie snelheid verhoogd . Voor het onderzoek naar de snelheid waarmee geadsorbeerd $Fe2+$ en $Mn2+$ en het effect van de pH-waarde hierop, zijn kortdurende experimenten uitgevoerd met kleine filter kolommen. "Met ijzer gecoat zand" en Aquamandix zijn beladen met $Fe2+$ en $Mn2+$ onder zuurstofloze condities, vervolgens werd zuurstofhoudend water door de kolom geleid en het zuurstof gehalte in het effluent gemeten. De zuurstof die werd verbruikt door het geadsorbeerde ferro op "met het ijzer gecoat filterzand" was bij een pH-waarde van 8 ongeveer 4 maal groter dan bij een pH-waarde van 6, hierbij werd respectievelijk 8.6% en 29.2% van het geadsorbeerd ferro geoxideerd. Het werkelijke verschil in de oxidatie snelheid is waarschijnlijk veel groter omdat het "met ijzer gecoat filterzand" de pH-waarde bufferde tijdens de oxidatie test door het oplossen van calcium carbonaat en mogelijk in de coating aanwezig ferro carbonaat.. Echter het aan Aquamandix geadsorbeerde $Mn2$ liet geen enkele opname van zuurstof zien tijdens de duur van de experimenten. Dit onverwachte resultaat geeft aan dat adsorptie van $Mn2+$ aan Aquamandix geen garantie is dat het ook snel wordt geoxideerd.

# List of symbols

units

| | | |
|---|---|---|
| $k_o$ | manganese oxidation rate constant | $l^2/mol^2.atm.min$ |
| $k_1$ | heterogenous manganese oxidation rate constant | $l^3/mol^3.atm.min$ |
| K | the apparent first order reaction rate constant | $min^{-1}$ |
| $k_{app}$ | overall Fe(II) oxidation rate constant | $l/mol. min^{-1}$ |
| $k^{\theta}$ | homogenous Fe(II) oxidation rate constant | $s^{-1}$ |
| $k^1$ | heterogeneous Fe(II) oxidation rate constant | $mg^{-1} s^{-1}$ |
| Ko | rate constant for Fe(II) oxidation | $l^3 mol^{-3} s^{-1}$ |
| $k_{so}$ | real rate constant for Fe (II) oxidation | $l\ mol^{-1} s^{-1}$ |
| $K_e$ | equilibrium constant | $mol\ l^{-1} mg^{-1}$ |
| ε(Ur) | total potential energy | $KJ\ mol^{-1}$ |
| C | dispersion | |
| r | distance separating two atoms | |
| B | empirical constant | |
| $P_{zc}$ | point of zero charge | |
| $q_s$ | amount of adsorbate adsorbed per unit mass (at equilibrium) | $mg/g$ or $g/m^2$ |
| Cs | equilibrium constant | $mg/l$ or $g/m^3$ or $g/l$ |
| c | liquid phase adsorbate concentration at time t | $kg/m^3$ |
| $c_o$ | initial liquid phase adsorbate concentration | $kg/m^3$ |
| $c_s$ | liquid phase adsorbate concentration near the particle surface | $kg/m^3$ |
| $k_2$ | rate constant for the adsorption process (Lagergren's equation) | |
| s | the number of active sites on the adsorbent occupied (PDSOK) | $mol/g$ |
| $s_s$ | the number of active sites on the adsorbent occupied (PDSOK) | |
| $k_3$ | the rate constant (PDSOK) | |
| a | the adsorbent dosage | $g/l$ |
| t | time | hours |
| $k_1$ | the LDF kinetic rate constant | $s^{-1}$ |
| $q_m$ | the maximum sorption capacity – Langmuir adsorption | $mg/g$ |
| b | the Langmuir adsorption constant | $l/mg$ |
| $q_e$ | the amount of adsorbate adsorbed at equilibrium per unit mass of adsorbent - Largergren model | $mol/g$ |
| $q_t$ | the amount of adsorbate adsorbed at any time t per unit mass of adsorbent - Largergren model | $mol/g$ |
| $X_m$ | the maximum sorption capacity - DKR model | |
| β | activity coefficient | |
| ε | Polanyi potential | |
| E | sorption energy – DKR model | KJ |
| $U_c$ | uniformity coefficient | |

# Abbreviations

| | |
|---|---|
| WWF | World Wide Fund for Nature |
| UNEP | United Nations Environment Program |
| WHO | World Health Organization |
| UNWWD | United Nations World Water Development |
| GDP | Gross Domestic Product |
| GWSC | Ghana Water and Sewage Co-operation |
| GWCL | Ghana Water Company Limited |
| CWSA | Communnity Water and Sanitation Agency |
| CSIR | Council for Scientific and Industrial Research |
| EAWAG | Swiss Federal Institute for Environmental Science and Technology |
| USEPA | United States Environmental Protection Agency |
| BAK | Bosomtwi Atwima Kwawoma |
| AAS | Ahafo Ano South |
| IOCS | Iron Oxide Coated Sand |
| $\equiv$S-OH | Neutral Surface hydroxyl group |
| $\equiv$S-OH$_2^+$ | Positively charged Surface hydroxyl group |
| $\equiv$S-O$^-$ | Negatively charged Surface hydroxyl group |
| A$_2^-$ | Hypothetical divalent anion |
| LDF | Linear Driving Force |
| PDSOK | Potential Driving Second Order Kinetic model |
| DKR | Dubinin-Kaganer-Radushkevisch |
| MGS | Manganese green sand |
| LmIOCS | Low Manganese Iron Oxide Coated Sand |
| AQM | Aquamandix |
| RMGS | Regenerated Manganese green sand |
| GWTP | Groundwater treatment plant |
| MOCS | Manganese Oxide Coated Sand |
| MW | Media washing |
| LS | Limestone layer |
| Rd | Regular draining |
| IOCP | Iron Oxide Coated Pumice |
| Dv | Drain valve |
| ARV | Air release valve |
| V | Valve |
| OXN | Oxidation |
| BLK | Blank |
| Inf.DO | Influent Dissolved Oxygen |
| LAT | Laterite |

# List of publications and presentations

1) **Buamah, R**. Petrusevski, B. and Schippers, J.C. (2008) Adsorptive removal of manganese (II) from the aqueous phase using iron oxide coated sand. *Journal of Water Supply: Research and Technology – AQUA* 57.1:1 – 12.

2) **Buamah, R**. Petrusevski, B. and Schippers, J.C. (2008) Presence of arsenic, iron and manganese in groundwater within the gold-belt zone of Ghana. *Journal of Water Supply: Research and Technology – AQUA* 57.7:519 – 529.

3) **Buamah, R**. Petrusevski, B. de Ridder, D. van de Wetering, T.S.C.M. and Schippers, J.C. (2009) Manganese removal in groundwater treatment: practice, problems and probable solutions. *Journal of Water Science and Technology: Water Supply* 9.1: 89 – 98.

4) **Buamah, R**. Petrusevski, B. and Schippers, J.C. (2009) The influence of process conditions on the oxidation kinetics of adsorbed ferrous. *Journal of Water Science and Technology: Water Supply* (in press).

5) **Buamah, R**. Petrusevski, B. and Schippers, J.C. (2008) Manganese (II) adsorption characteristics of selected filter media for groundwater treatment: equilibrium and kinetics. In: *Proceedings of the IWA World Water Congress* at Vienna, 7 – 12 September, 2008 pp 149 (given as oral presentation and a paper).

6) **Buamah, R**. Petrusevski, B. de Ridder, D. van de Watering and Schippers, J.C. (2008) Manganese removal in groundwater treatment: practice, problems and probable solutions. In: *Proceedings of the International Water Association (IWA) 5th Leading Edge Conference on Water and Waste Technologies*. (Nutrient Removal), June 1 - 4, 2008; Zurich – Switzerland (presented as a poster and a paper).

7) **Buamah R**. Petrusevski, B. and Schippers, J.C. (2008) Oxidation of adsorbed ferrous and manganese ions: kinetics and Influence of process conditions. In: *Proceedings of the University Council of Water Resources / National Institutes for Water Resources Annual Conference on International Water Resources*: Challenges for the 21st Century. Durham, North Carolina; July 22 – 24, 2008 (given as oral presentation and a paper).

8) Petrusevski, B., **Buamah, R**., Sharma, S., Krius, F., Slokkar, Y. and Schippers, J.C. (2008) Adsorptive removal of arsenic, iron, manganese and chromium with iron oxide coated sand. In: *Proceedings of the 1st International Conference G16 Research Frontiers in Chalocogen Cycle Science and Technology*, Wageningen, The Netherlands.

9) **Buamah R**. Petrusevski, B and Schippers, J.C. (2002) Presence and Health related effects of arsenic, iron and manganese in groundwater in Ashanti and Western regions of Ghana. (presented at the CEMSA conference – South Africa). In *Proceedings of the CEMSA conference South Africa* – September 2002.

# Curriculum Vitae

Richard Buamah was born on 18th December 1966 at Bantama, Kumasi – Ghana. In 1985, he completed his high school education at Opoku Ware Senior High School and proceeded to pursue a Bachelor of Science degree in Biochemistry at the Kwame Nkrumah University of Science and Technology (KNUST). In 1990, he graduated with a Bachelor of Science degree, second class honours – upper division and followed it up with a national service at KNUST.

In 1995, he completed his Master of philosophy degree programme in Biochemistry at KNUST. In 1997, he was employed as an analyst and lecturer at KNUST working specifically with Water Resources and Environmental Sanitation Project of the Civil Engineering Department at KNUST. In October 2000, he joined the UNESCO-IHE, Institute for Water Education (the Urban Water and Infracture department) to pursue a number of graduate courses in Water Treatment and followed it up with a PhD programme under sandwinch construction.

He has twelve publications to his credit and has presented papers and posters at several peer reviewed conferences including the Biennial World Water Congress of the IWA (Sept 7 – 12, 2008 at Vienna, Austria), the Leading Edge Technology Conference of the IWA (June 4 – 6, 2008 at Zurich, Switzerland) and UCOWR/NIWR conference, (July 22 – 24, at Durham, North Carolina – USA) etc.

T - #0071 - 071024 - C72 - 234/156/11 - PB - 9780415573795 - Gloss Lamination